Community School Mathematics 3B

Pat Lilburn
Pam Rawson
Peter Sullivan

Oxford University Press
Department of Education, Papua New Guinea

OXFORD
UNIVERSITY PRESS

Oxford University Press is a department of the University of Oxford. It furthers the University's objective of excellence in research, scholarship, and education by publishing worldwide. Oxford is a registered trademark of Oxford University Press in the UK and in certain other countries.

Published in Australia by
Oxford University Press
253 Normanby Road, South Melbourne, Victoria 3205, Australia

and

The Department of Education, Papua New Guinea

First published 1993
Reprinted 1993, 2009, 2012 (D)

ISBN 978 9980 5885792

Written by Pat Lilburn, Pam Rawson and Peter Sullivan
Writing consultants Elsie Kinavai and Kate Deutrom
Papua New Guinea project co-ordinator and
writing consultant Katherine Schneider
Cover photograph by Rocky Roe Photographics
Illustrated by Annie Vanston and Mary Ann Hurley
Typeset by Solo Typesetting, South Australia
Printed and bound in Australia by Ligare Book Printers, Pty Ltd

Secretary's Message

This pupil's book is part of a new Mathematics Program designed and written for use during the early years of schooling in Papua New Guinea. The core materials consist of two pupil books called **Community School Mathematics 3A** and **Community School Mathematics 3B**. They are accompanied by the **Teacher's Resource Book 3**. They will replace the MACS series now in use.

In the Community School Mathematics Program children are taught mathematics by first using real objects. Later, the children will use pictures of objects. Finally, the children will use number symbols to represent these objects. Each school will be supplied with a set of **base ten blocks** to use during mathematics lessons. You and your pupils must also collect other real objects such as seeds, nuts, shells etc. **Always allow your pupils to use real objects during their mathematics lessons if they want to. The children will decide when they do not need the help of real objects any longer**. Remember, when children use real objects, they will understand mathematics better.

This program also encourages the children to solve their own problems and make their own decisions with confidence. In order to learn these skills, the children should talk about what they are doing in every lesson. **Since learning is most effective when it has meaning and is enjoyable, allow the children to use the language that they are most comfortable with**. However, in Grade 3, the main language of instruction should be English, with other languages including the vernacular, used to aid understanding of new and difficult concepts.

Finally, the National Department of Education wants teachers to be **flexible** in programming and timetabling. This book will show you some ways to do this.

J. E. Tetaga OBE
Secretary for Education

Contents

Unit 11 *Travel in Papua New Guinea*

Unit 12 *Revision and Investigation*

30 toea

15 toea

52 toea

3 toea

40 toea

7 toea

11 toea

25 toea

22 toea

6 toea

25 toea

80 toea

32 toea

60 toea

55 toea

75 toea

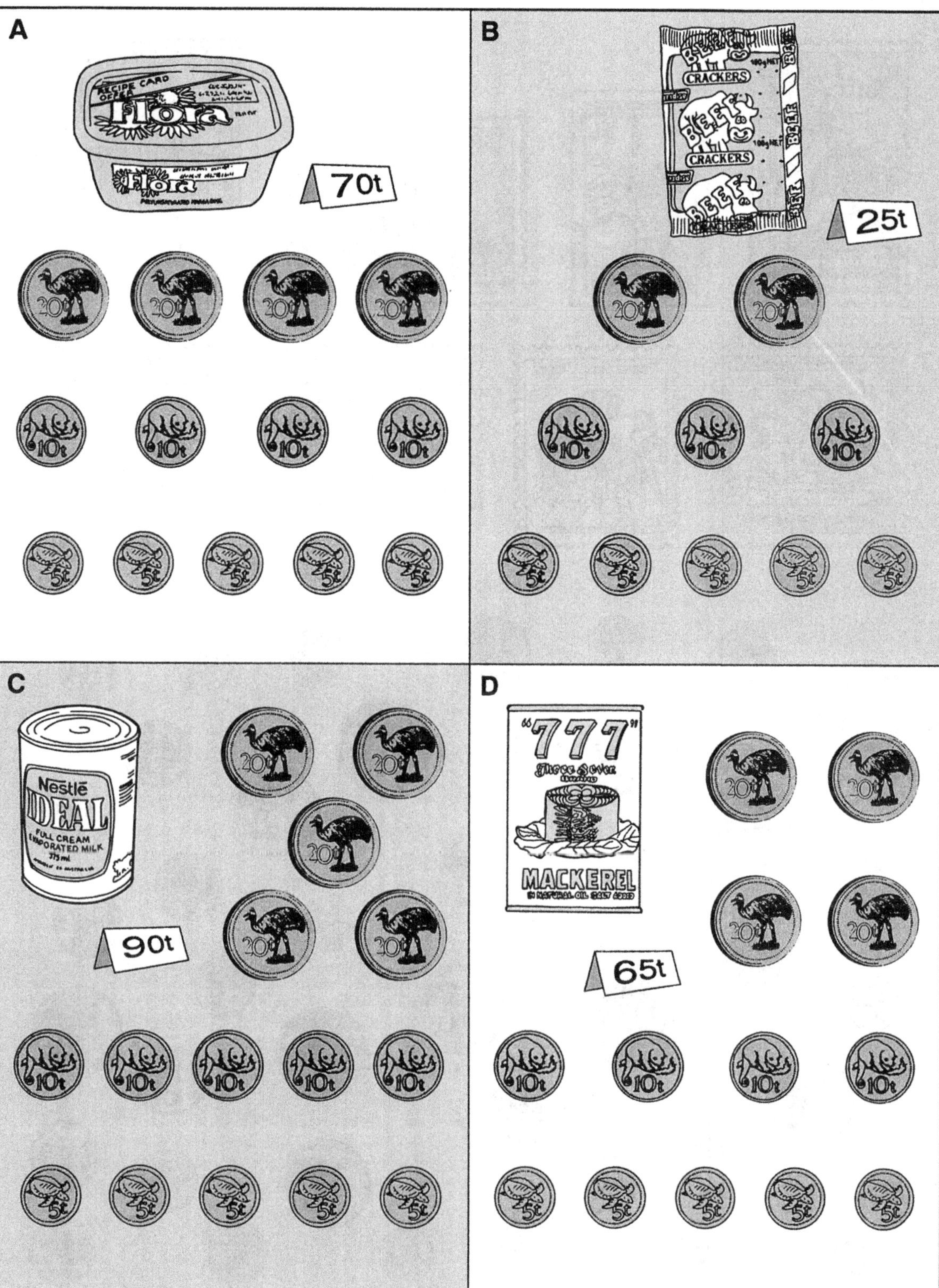
A
RECIPE CARD OFFER
Flora
Flora
70t
B
BEEF
CRACKERS
BEEF
CRACKERS
BEEF
25t
C
Nestlé
IDEAL
FULL CREAM
EVAPORATED MILK
90t
D
"777"
MACKEREL
65t
20t
10t
5t

35t
papua new guinea

GOGODALA DANCE MASKS
20t
PAPUA NEW GUINEA

Papua New Guinea
BASILAKI MILNE BAY
70t

LAUAN NEW IRELAND
60t
Papua New Guinea

1t
Papua New Guinea

Papua New Guinea
3t

21t
PAPUA NEW GUINEA

30t
Papua New Guinea

You buy	You give	How much change?
A 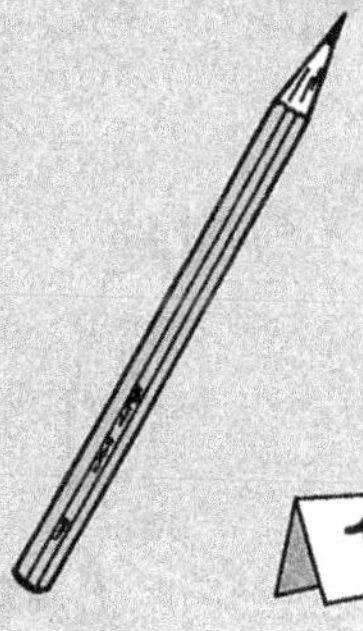 14t	 20t	
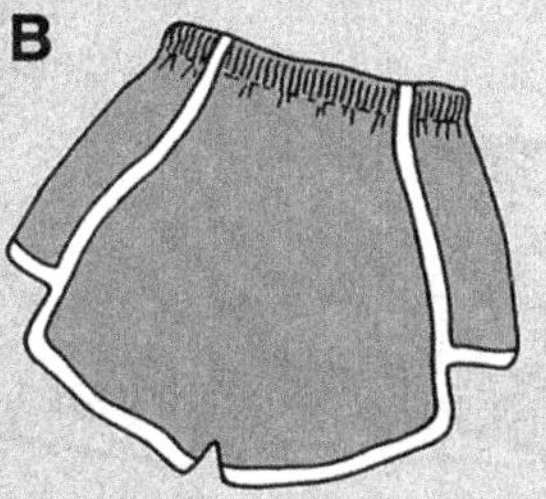 B 50t	 20t 20t 20t	
 C 83t	 50t 20t 20t	
 D 36t	 50t	

A

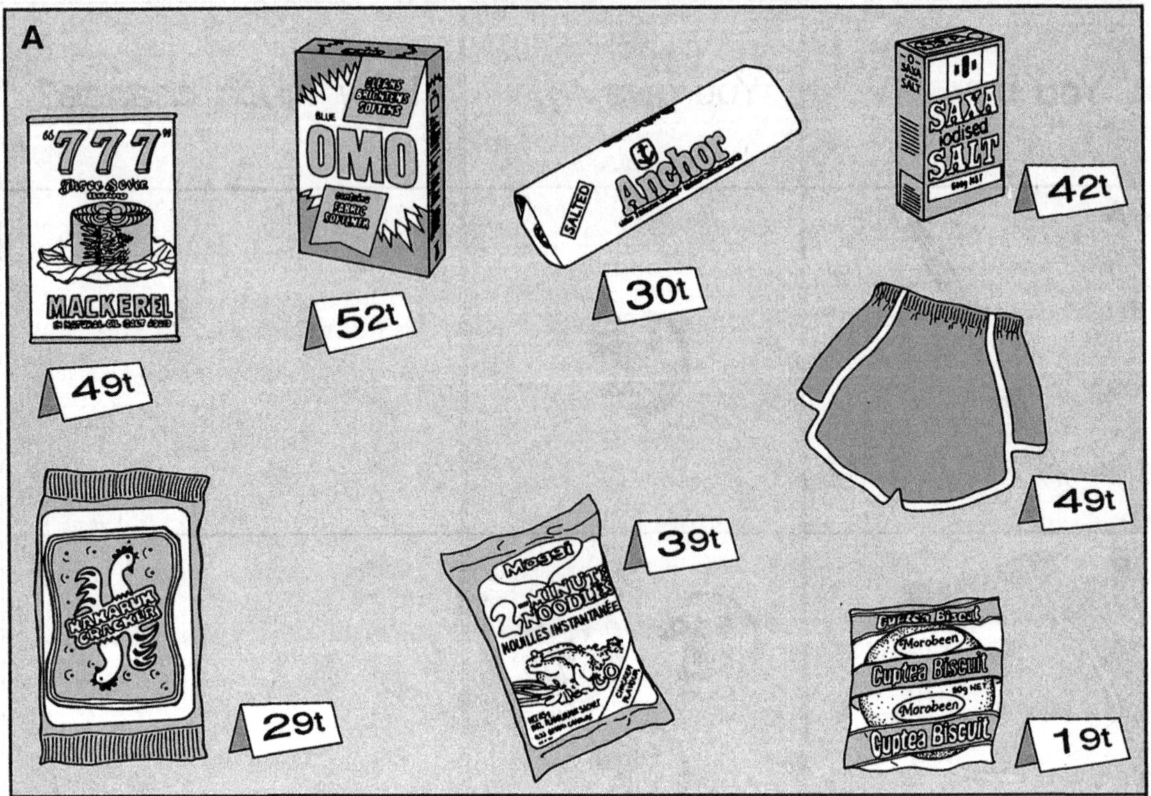

49t

52t

30t

42t

49t

29t

39t

19t

B

You buy:

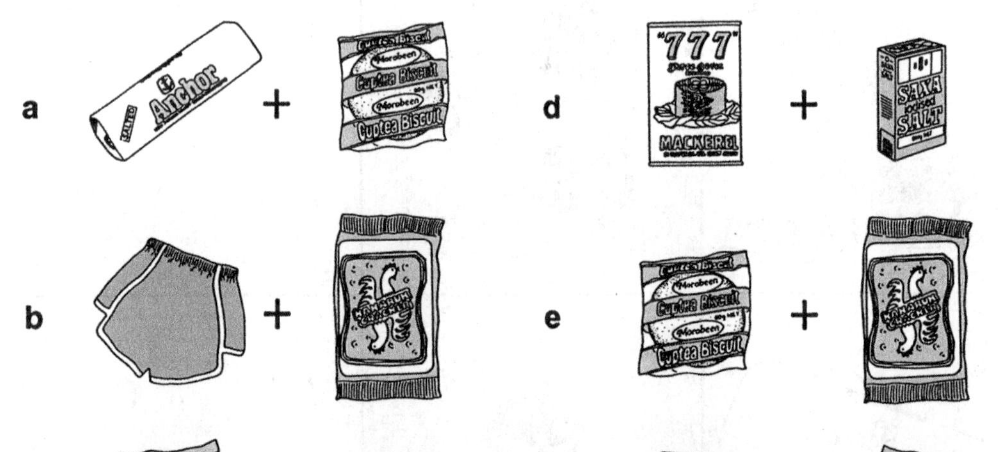

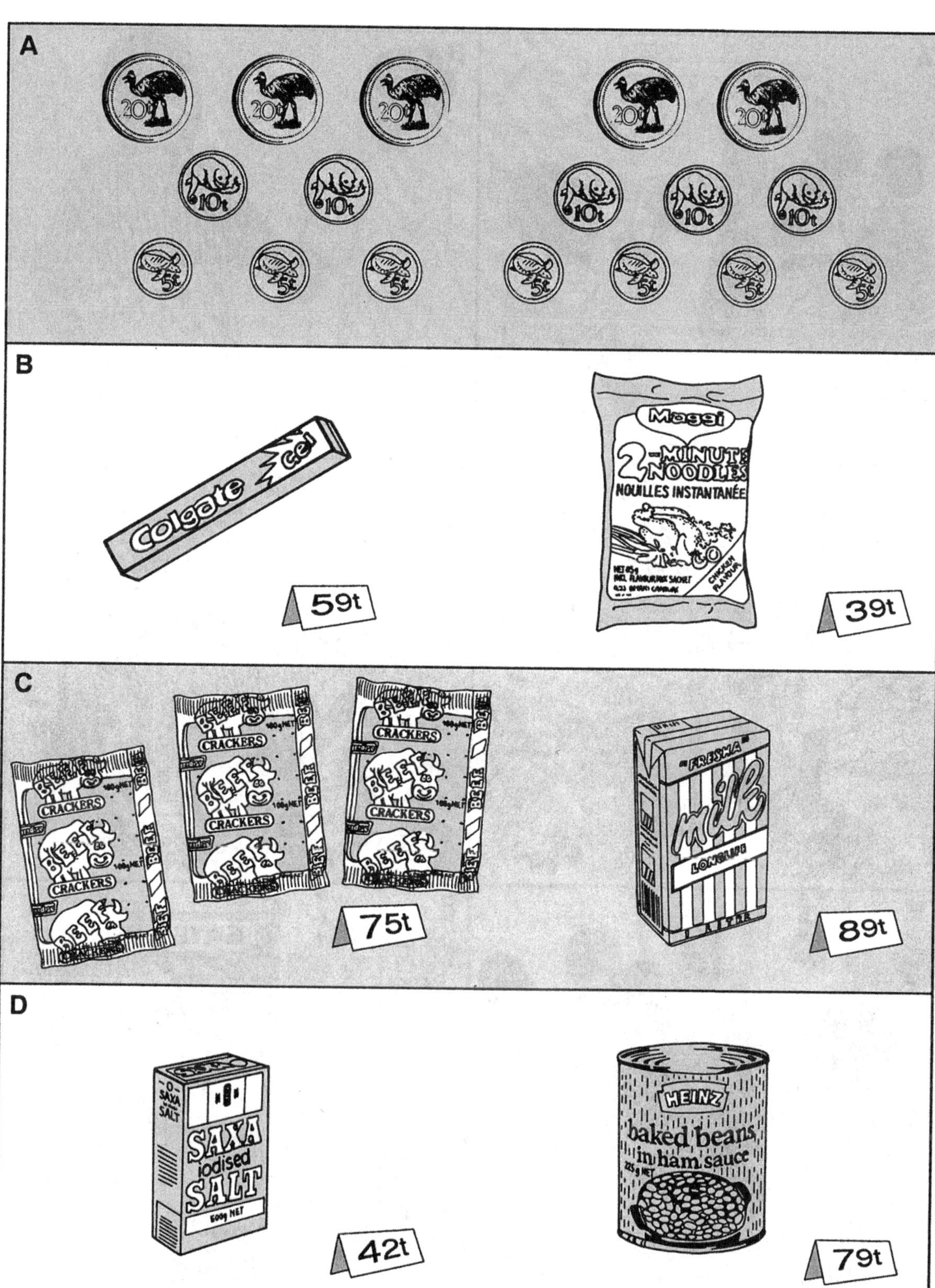
A
20t
20t
20t
20t
20t
10t
10t
10t
10t
10t
5t
5t
5t
5t
5t
5t
5t
B
Colgate gel
59t
Maggi
2-MINUTE NOODLES
NOUILLES INSTANTANÉE
CHICKEN FLAVOUR
39t
C
BEEF
CRACKERS
75t
FRESHA
milk
LONGLIFE
89t
D
SAXA
iodised
SALT
600g NET
42t
HEINZ
baked beans
in ham sauce
225 g NET
79t

A
B
Post-Co
may
C
D
E
10
ten
F
GATE 2
THE BIG GAME
THE BIG GAME

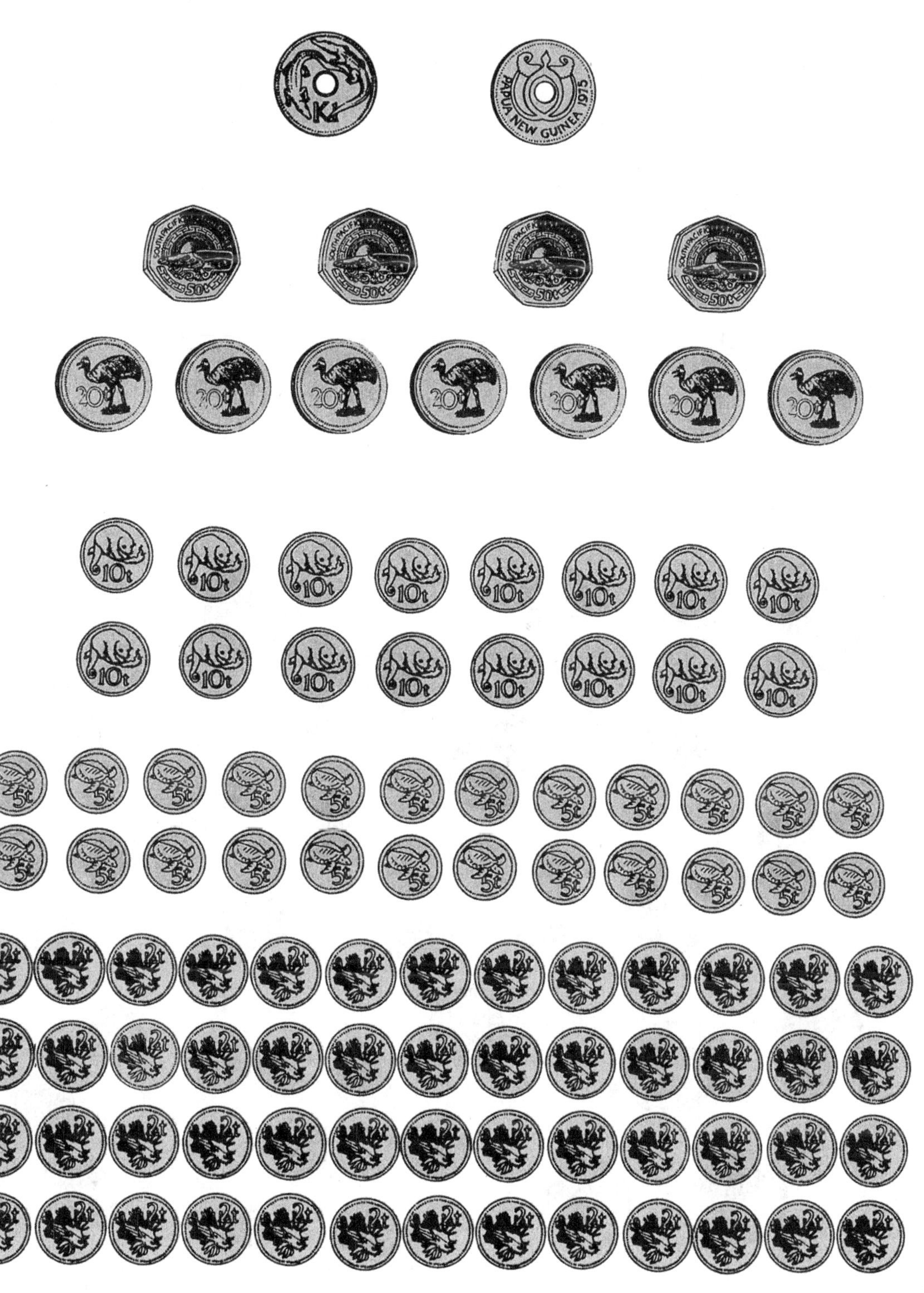
PAPUA NEW GUINEA 1975
50t
20t
10t
5t
2t

2 kina 10 toea

1 kina 30 toea

K1 55t

K2 60t

K1 35t

K1 77t

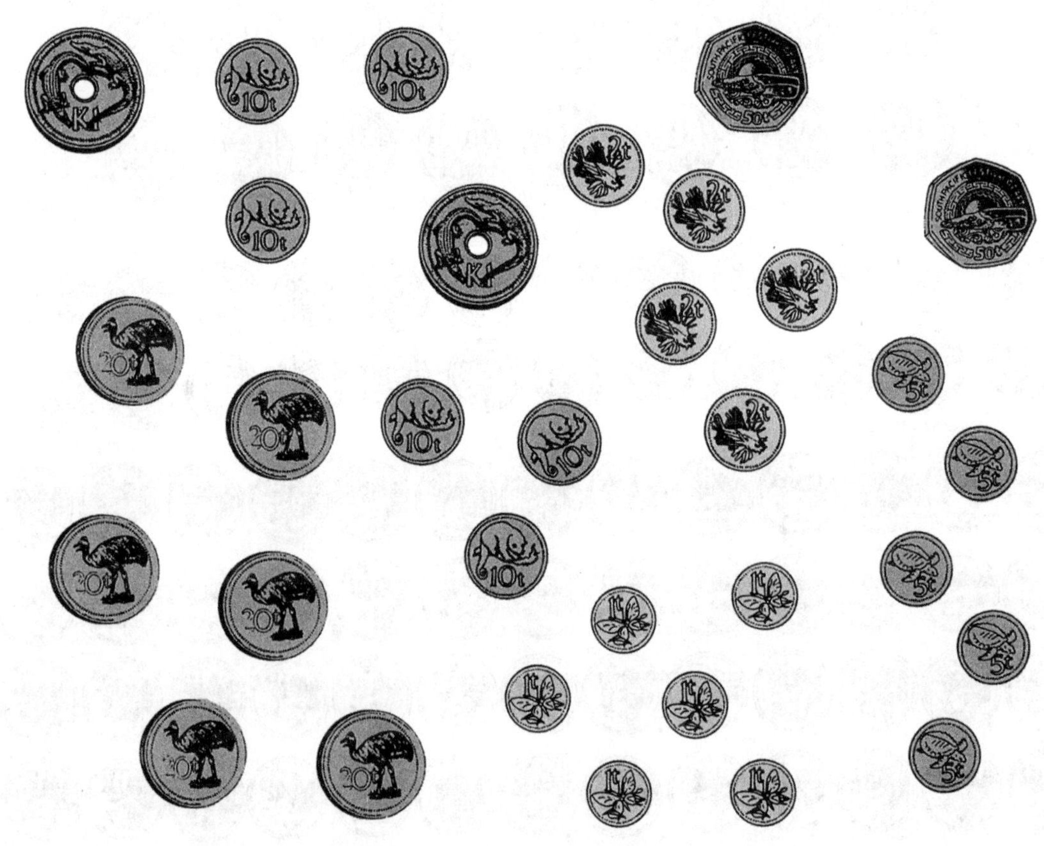

You buy	You give	How much change?
A 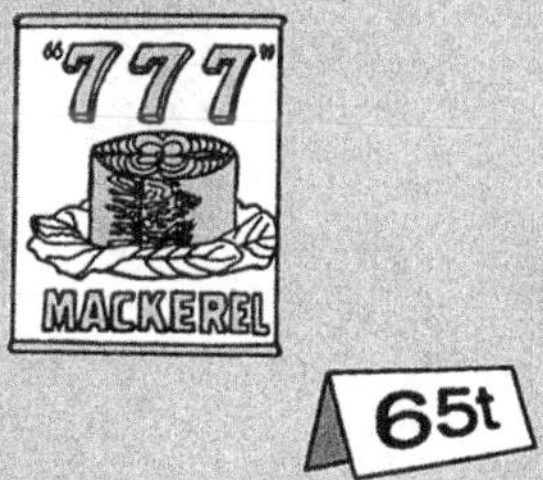 "777" MACKEREL 65t	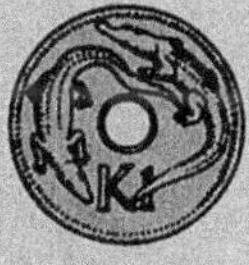 K1	
B 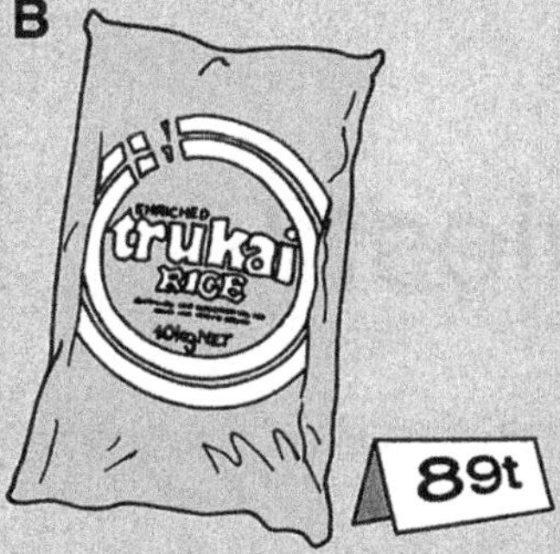 trukai RICE 89t	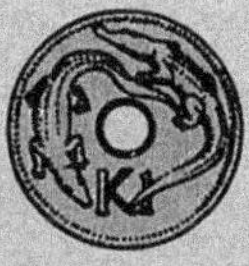 K1	
C 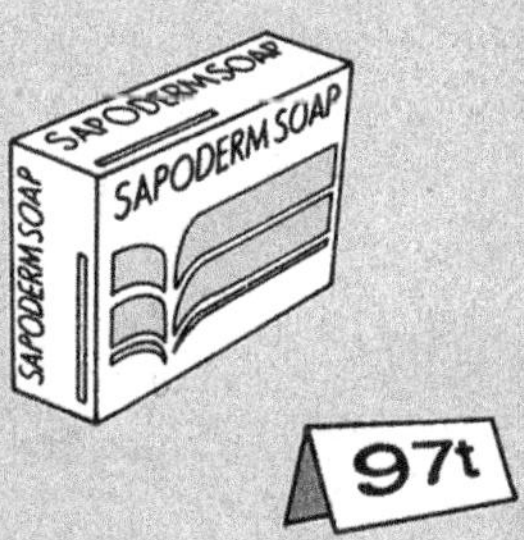 SAPODERM SOAP 97t	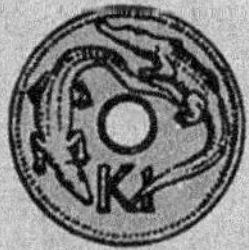 K1	
D 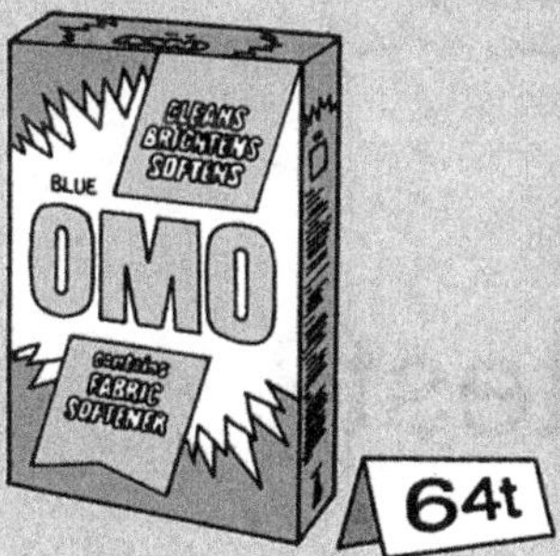 BLUE OMO CLEANS BRIGHTENS SOFTENS Contains FABRIC SOFTENER 64t	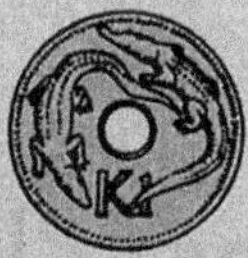 K1	

Which is the heaviest?

B

Which is the lightest?

ENRICHED
trukai
RICE
1kg NET
trukai

Flame
Plain Flour
1kg NET

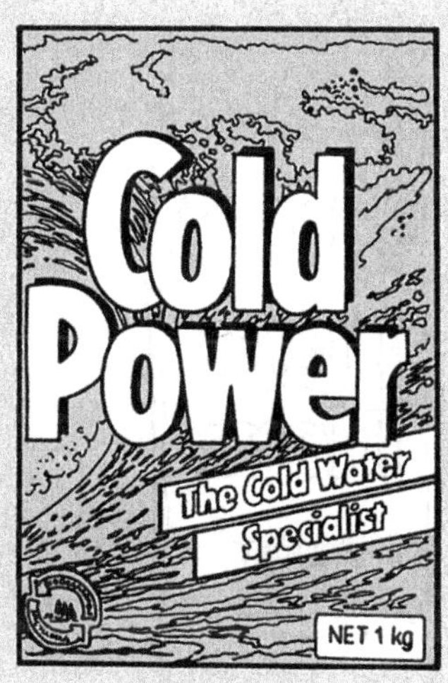
Cold Power
The Cold Water Specialist
NET 1 kg

MOISTURE PROOF
Kooka
IODISED COOKING SALT
1kg NET
Kooka COOKING SALT

RAMU
RAMU
RAMU
SUGAR
1kg nett

Nestlé
Sunshine instant
INSTANT FULL CREAM MILK POWDER
NET 1kg MADE IN AUSTRALIA

15kg
Nestle
MILO
Food Drink
Net 1·25kg
Flame
Plain Flour
5kg NET
ENRICHED
trukai
RICE
10kg NET
CLEANS
BRIGHTENS
SOFTENS
BLUE
OMO
contains
FABRIC
SOFTENER
1·5kg

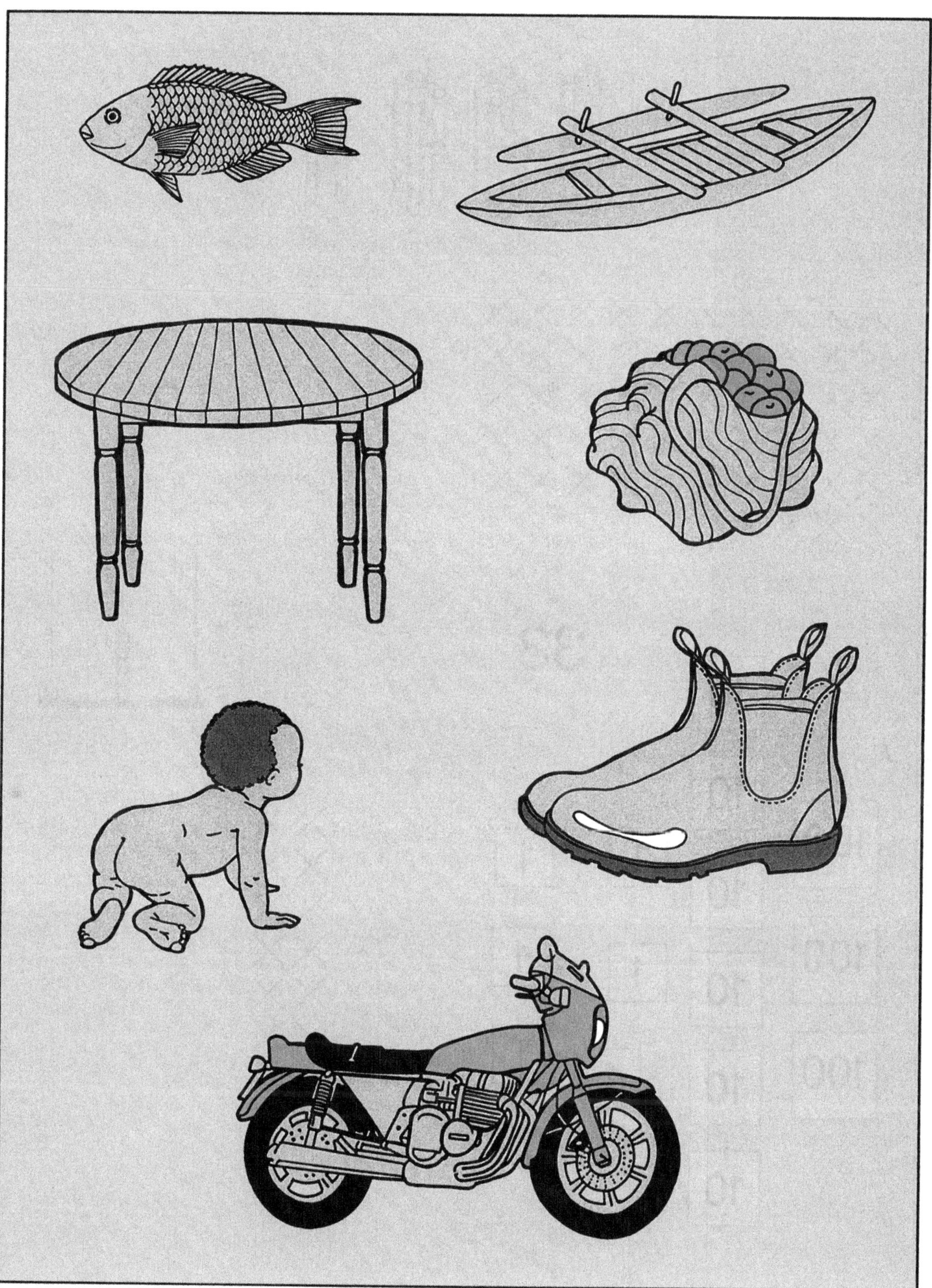

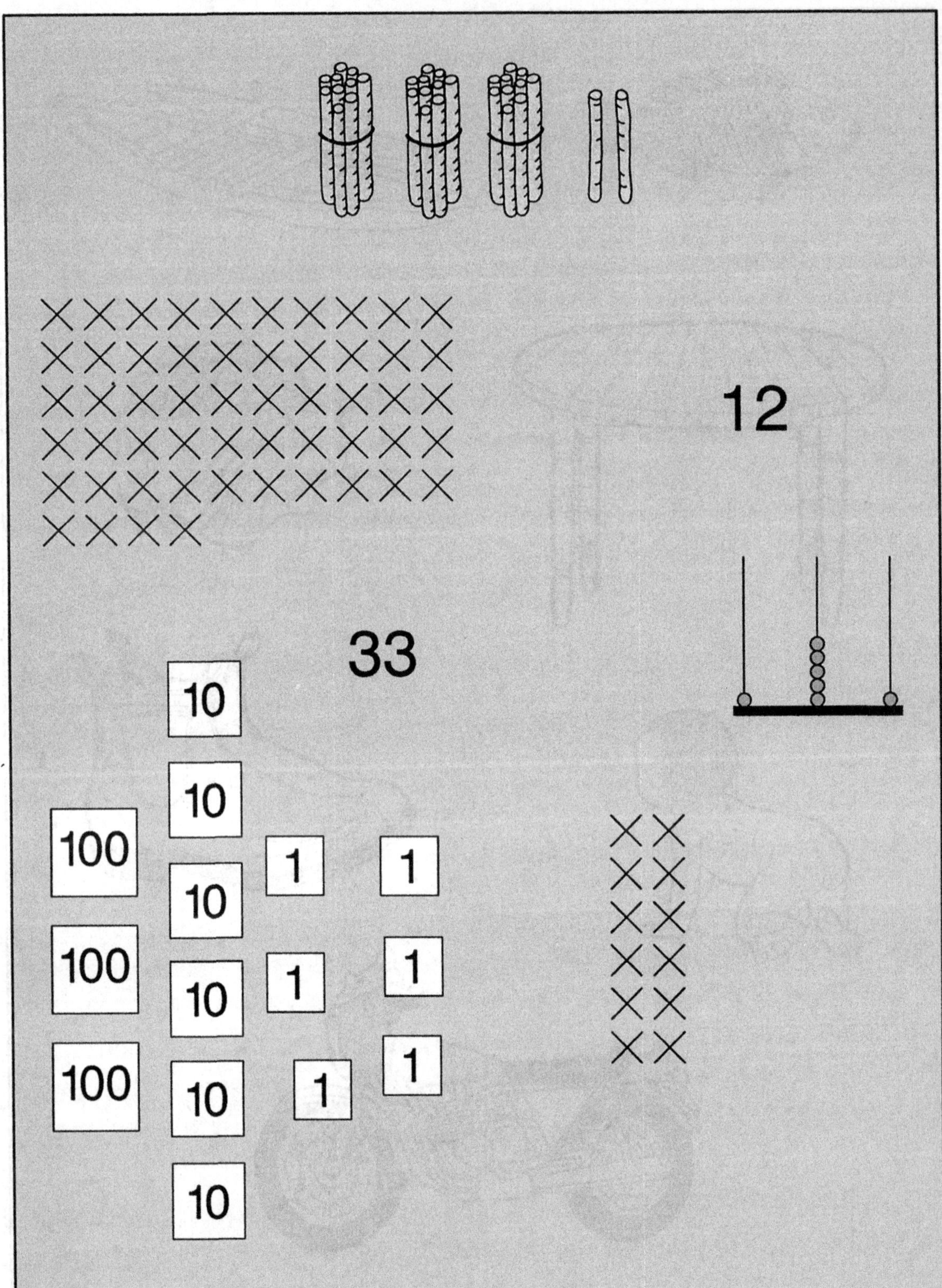
12
33
100
100
100
10
10
10
10
10
10
1
1
1
1
1
1

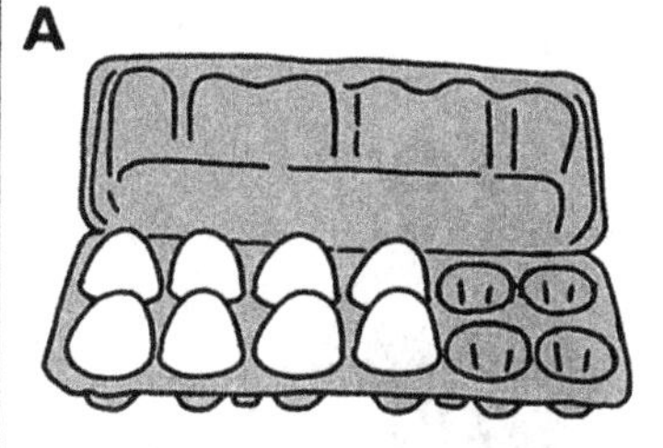
A

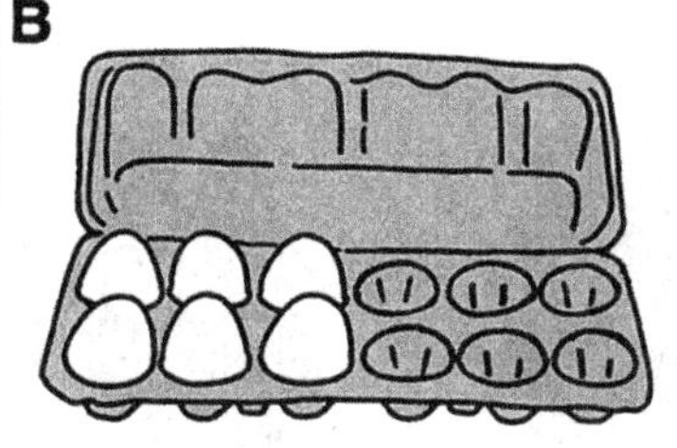
B

C

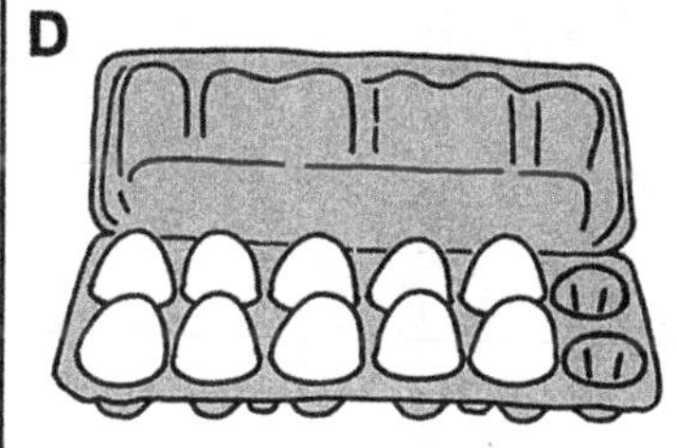
D

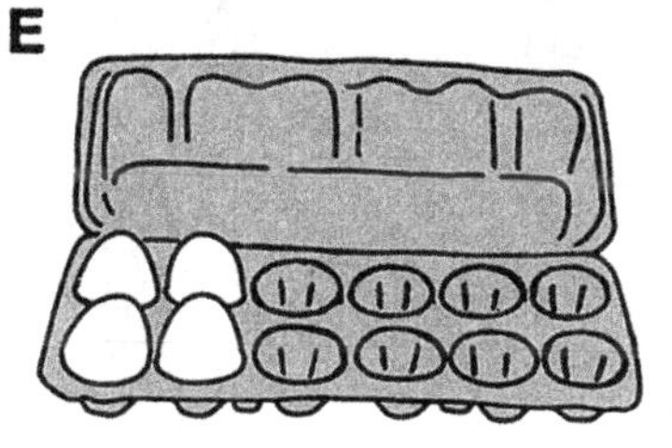
E

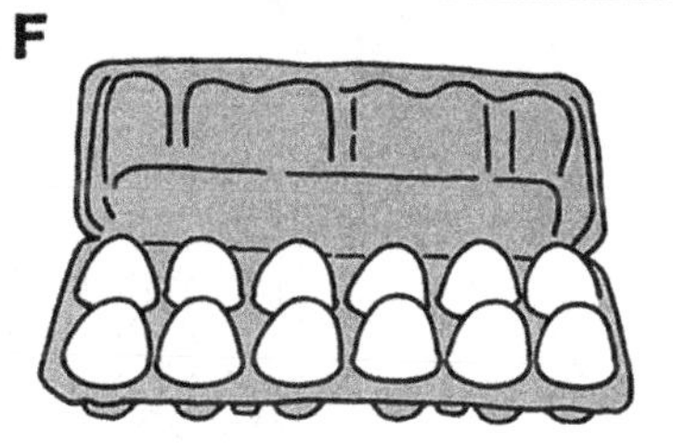
F

$1 \times 2 = 2$
$2 \times 2 = 4$
$3 \times 2 = 6$
$4 \times 2 = 8$
$5 \times 2 = 10$
$6 \times 2 = 12$
$7 \times 2 = 14$
$8 \times 2 = 16$
$9 \times 2 = 18$
$10 \times 2 = 20$

$1 \times 3 = 3$
$2 \times 3 = 6$
$3 \times 3 = 9$
$4 \times 3 = 12$
$5 \times 3 = 15$
$6 \times 3 = 18$
$7 \times 3 = 21$
$8 \times 3 = 24$
$9 \times 3 = 27$
$10 \times 3 = 30$

$1 \times 4 = 4$
$2 \times 4 = 8$
$3 \times 4 = 12$
$4 \times 4 = 16$
$5 \times 4 = 20$
$6 \times 4 = 24$
$7 \times 4 = 28$
$8 \times 4 = 32$
$9 \times 4 = 36$
$10 \times 4 = 40$

$1 \times 5 = 5$
$2 \times 5 = 10$
$3 \times 5 = 15$
$4 \times 5 = 20$
$5 \times 5 = 25$
$6 \times 5 = 30$
$7 \times 5 = 35$
$8 \times 5 = 40$
$9 \times 5 = 45$
$10 \times 5 = 50$

$1 \times 10 = 10$
$2 \times 10 = 20$
$3 \times 10 = 30$
$4 \times 10 = 40$
$5 \times 10 = 50$
$6 \times 10 = 60$
$7 \times 10 = 70$
$8 \times 10 = 80$
$9 \times 10 = 90$
$10 \times 10 = 100$

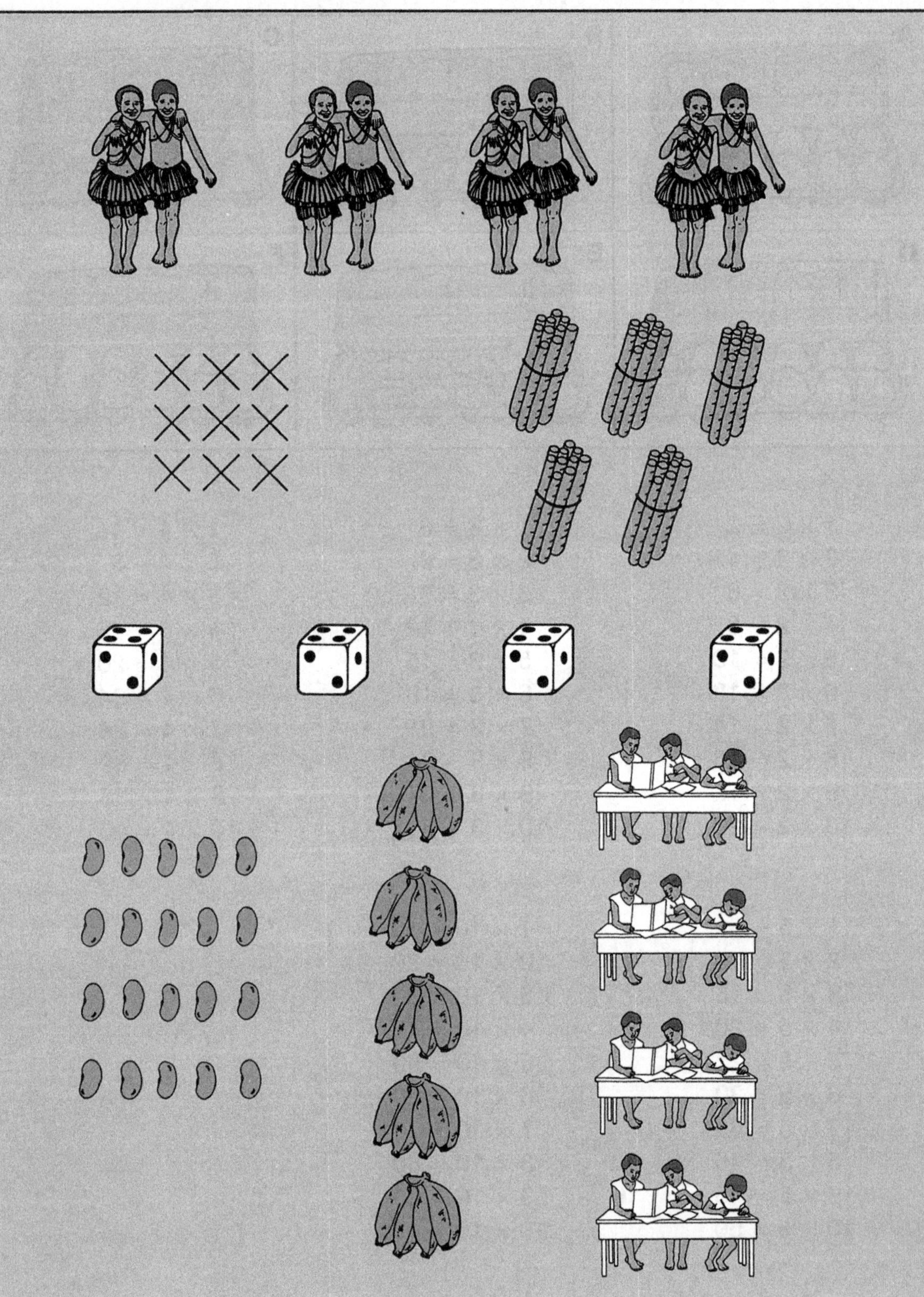

A

Make groups of 4.

B

Make groups of 10.

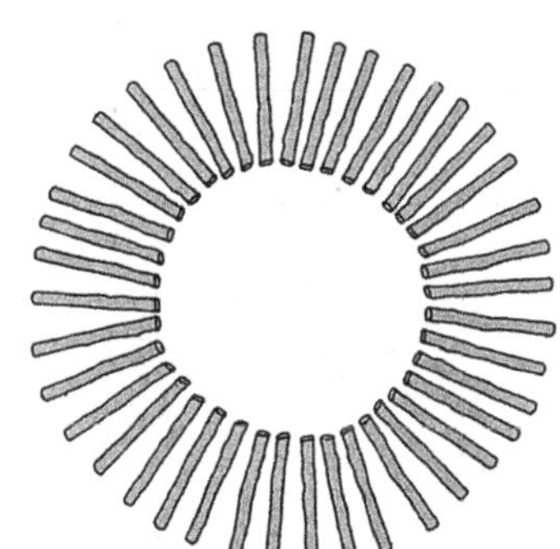

C

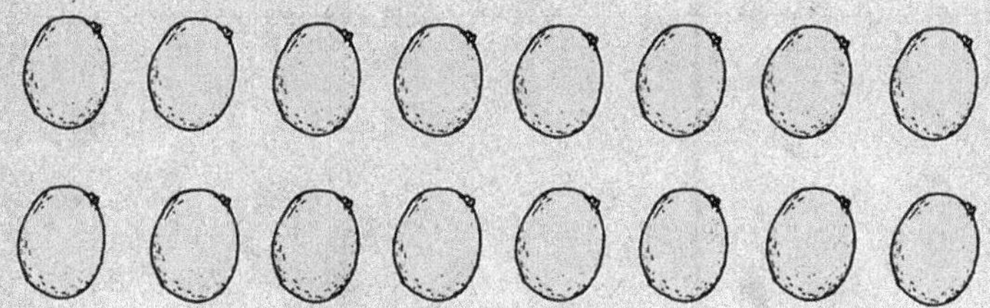

Share between 4.

D

Share between 3.

Before	After
A	
B	
C	
D	

a $4 \times 5 = 20$

b $3 \times 4 = 12$

c $3 \times 20 = 60$

d $6 \times 2 = 12$

e $5 \times 10 = 50$

f $9 \times 2 = 18$

g $20 \times 2 = 40$

A

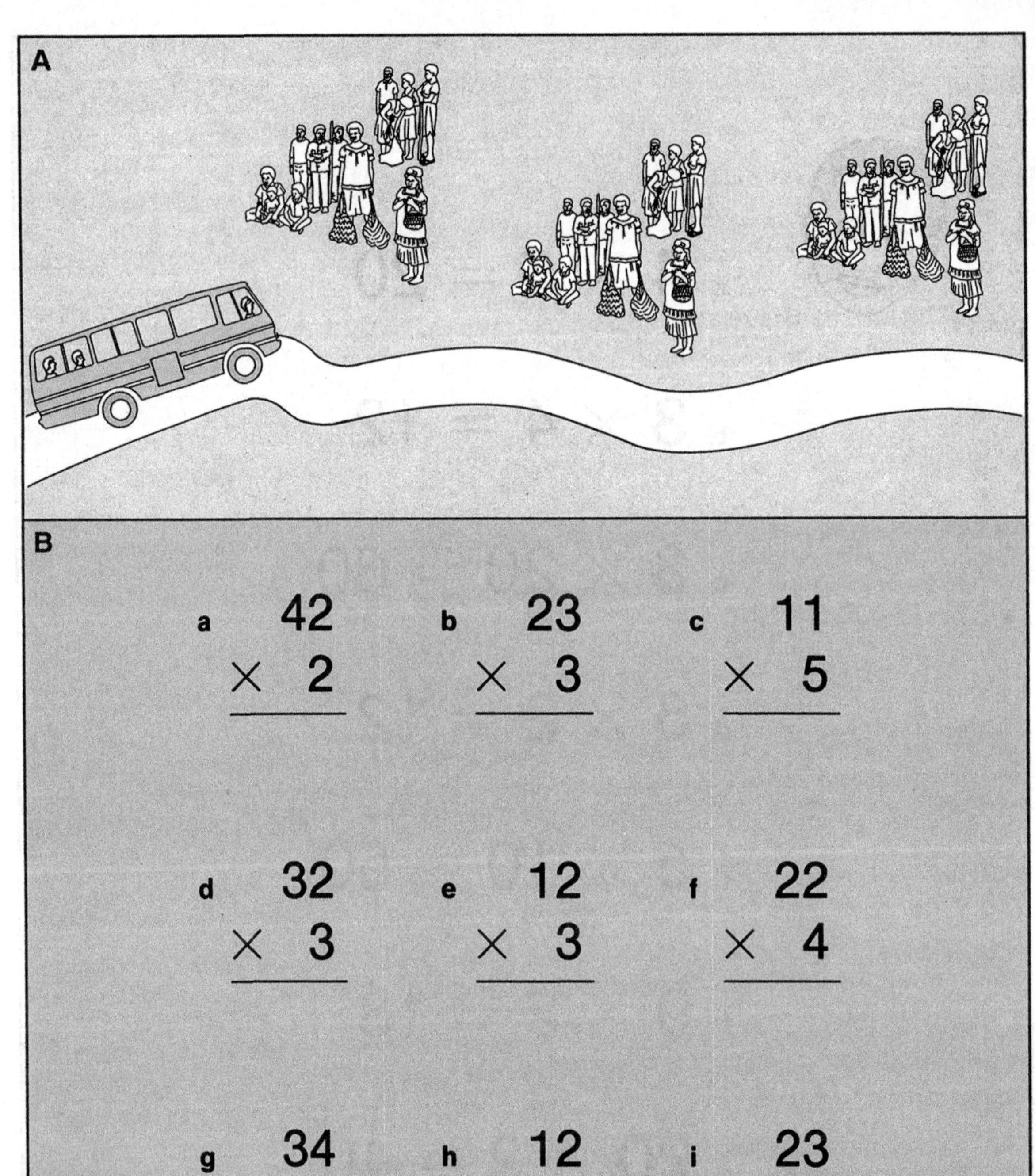

B

a $\begin{array}{r} 42 \\ \times \quad 2 \\ \hline \end{array}$

b $\begin{array}{r} 23 \\ \times \quad 3 \\ \hline \end{array}$

c $\begin{array}{r} 11 \\ \times \quad 5 \\ \hline \end{array}$

d $\begin{array}{r} 32 \\ \times \quad 3 \\ \hline \end{array}$

e $\begin{array}{r} 12 \\ \times \quad 3 \\ \hline \end{array}$

f $\begin{array}{r} 22 \\ \times \quad 4 \\ \hline \end{array}$

g $\begin{array}{r} 34 \\ \times \quad 2 \\ \hline \end{array}$

h $\begin{array}{r} 12 \\ \times \quad 4 \\ \hline \end{array}$

i $\begin{array}{r} 23 \\ \times \quad 2 \\ \hline \end{array}$

B

a
$$\begin{array}{r} 32 \\ \times \quad 2 \\ \hline \end{array}$$

b
$$\begin{array}{r} 21 \\ \times \quad 4 \\ \hline \end{array}$$

c
$$\begin{array}{r} 24 \\ \times \quad 2 \\ \hline \end{array}$$

d
$$\begin{array}{r} 31 \\ \times \quad 3 \\ \hline \end{array}$$

e
$$\begin{array}{r} 20 \\ \times \quad 4 \\ \hline \end{array}$$

f
$$\begin{array}{r} 43 \\ \times \quad 2 \\ \hline \end{array}$$

g
$$\begin{array}{r} 30 \\ \times \quad 3 \\ \hline \end{array}$$

h
$$\begin{array}{r} 22 \\ \times \quad 3 \\ \hline \end{array}$$

i
$$\begin{array}{r} 21 \\ \times \quad 3 \\ \hline \end{array}$$

A

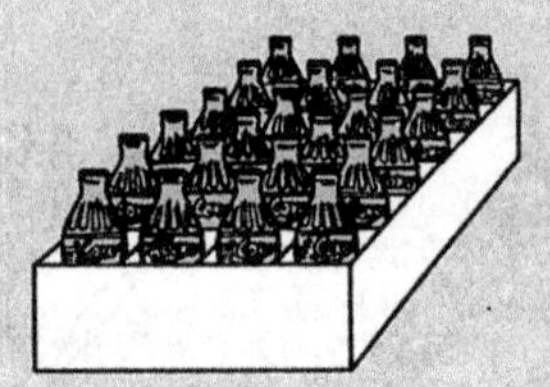

B

a $\begin{array}{r} 14 \\ \times \;\; 3 \\ \hline \end{array}$

b $\begin{array}{r} 16 \\ \times \;\; 4 \\ \hline \end{array}$

c $\begin{array}{r} 23 \\ \times \;\; 4 \\ \hline \end{array}$

d $\begin{array}{r} 15 \\ \times \;\; 3 \\ \hline \end{array}$

e $\begin{array}{r} 15 \\ \times \;\; 5 \\ \hline \end{array}$

f $\begin{array}{r} 12 \\ \times \;\; 5 \\ \hline \end{array}$

g $\begin{array}{r} 24 \\ \times \;\; 4 \\ \hline \end{array}$

h $\begin{array}{r} 13 \\ \times \;\; 5 \\ \hline \end{array}$

i $\begin{array}{r} 25 \\ \times \;\; 3 \\ \hline \end{array}$

96	42	75	69
70	48	36	30
120	92	40	150
60	56	140	230

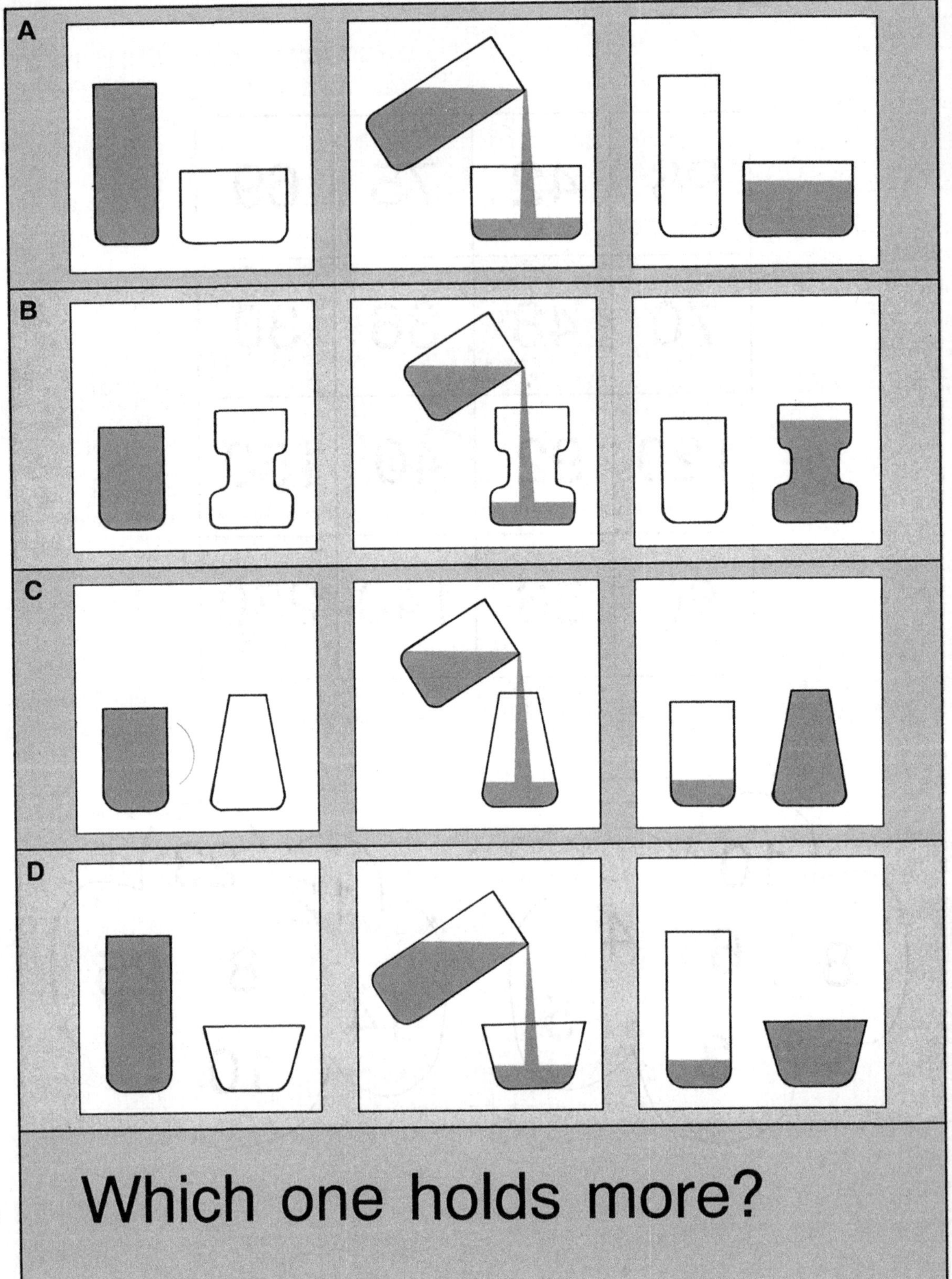
A
B
C
D
Which one holds more?

All these containers hold one litre.

These containers hold more than one litre.

These containers hold less than one litre.

A

B

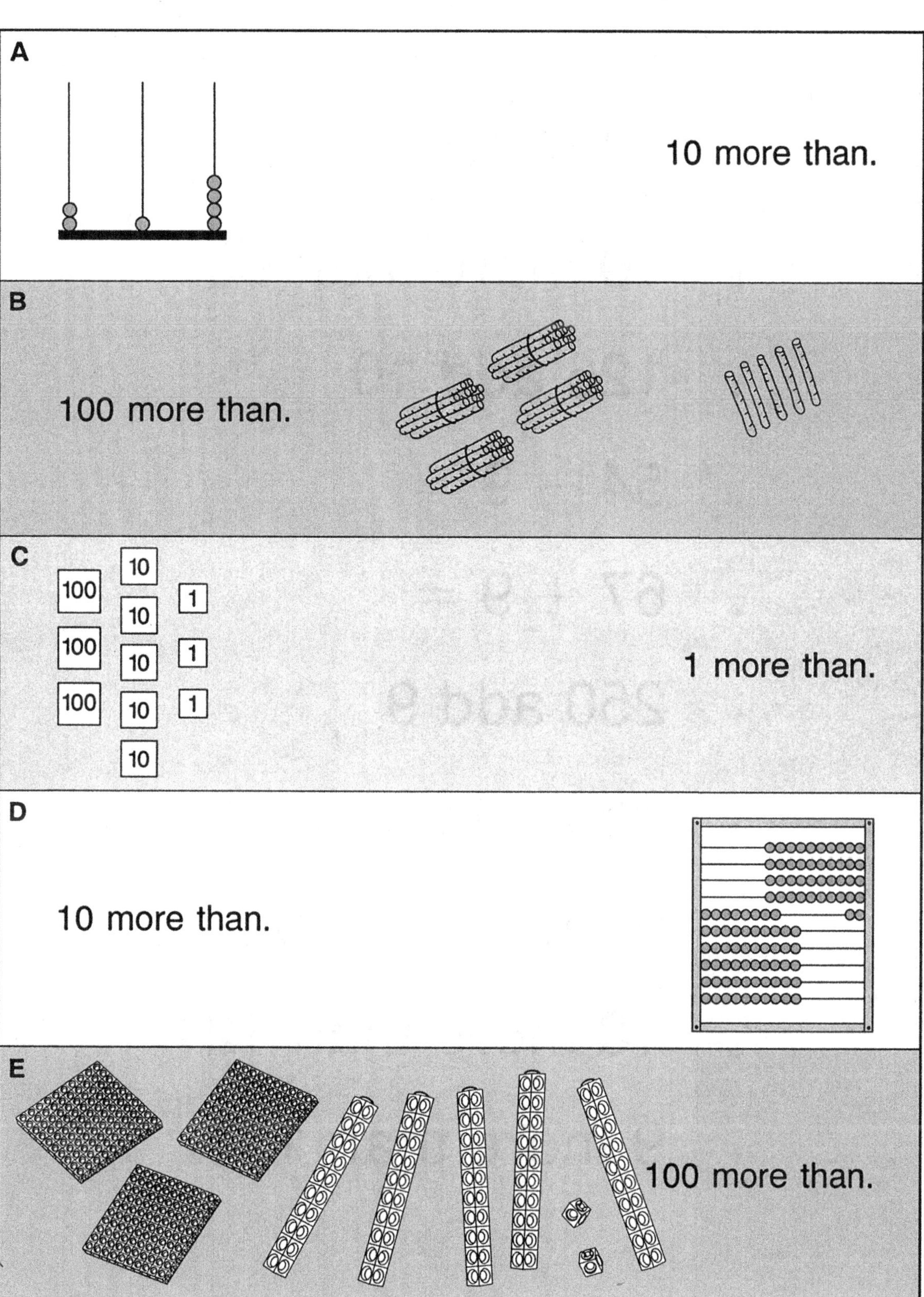
A
10 more than.
B
100 more than.
C
100
100
100
10
10
10
10
10
1
1
1
1 more than.
D
10 more than.
E
100 more than.

a $34 + 10 =$

b 10 more than 63

c 126 add 10

d $54 + 9 =$

e $67 + 9 =$

f 250 add 9

g 256 add 9

h 9 more than 52

i 10 more than 121

j 9 more than 121

a 34 — 10 =

b 52 take away 10

c 10 less than 251

d 10 less than 505

g 34 — 9 =

f 52 take away 9

g 9 less than 130

h 115 take away 9

i 200 take away 9

j 505 — 9 =

a $23 + 20 =$

b $56 + 20 =$

c $157 + 20 =$

d $120 + 20 =$

e $12 + 20 =$

f $34 + 20 =$

g $66 + 20 =$

h $152 + 20 =$

i $89 + 20 =$

j $180 + 20 =$

k $23 + 21 =$

l $56 + 21 =$

m $157 + 21 =$

n $240 + 21 =$

o $8 + 21 =$

p $15 + 21 =$

q $277 + 21 =$

r $187 + 21 =$

s $89 + 21 =$

t $252 + 21 =$

a $124 + 100 =$

b 100 more than 351

c 671 add 100

d $100 + 367 =$

e $350 + 99 =$

f 227 add 99

g 99 more than 88

h 260 add 99

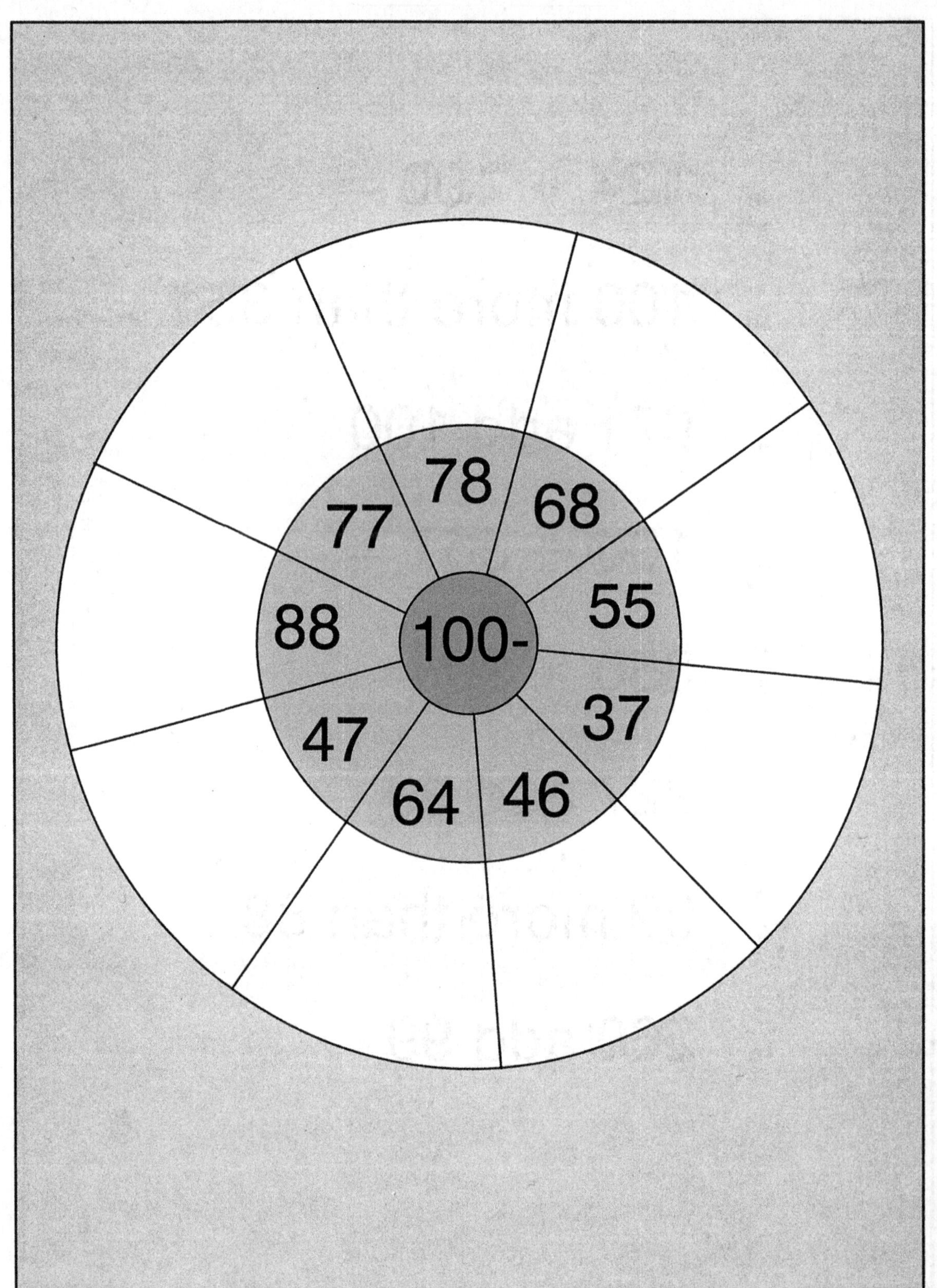
78
77
68
88
100-
55
37
47
64
46

How much is needed to buy:

2 x 

2 x 

2 x 

2 x 

2 x 

2 x 

A

What number am I?

- My hundreds number is more than 2 but less than 5.
- My tens number is less than my hundreds number.
- My units number is less than my tens number.
- All my digits are different.
- I am an even number.

B

What number am I?

- I am bigger than 150.
- My hundreds digit is 1.
- My tens number is more than my units number.
- You say my number when you count by 5's.
- My digits add up to 8.

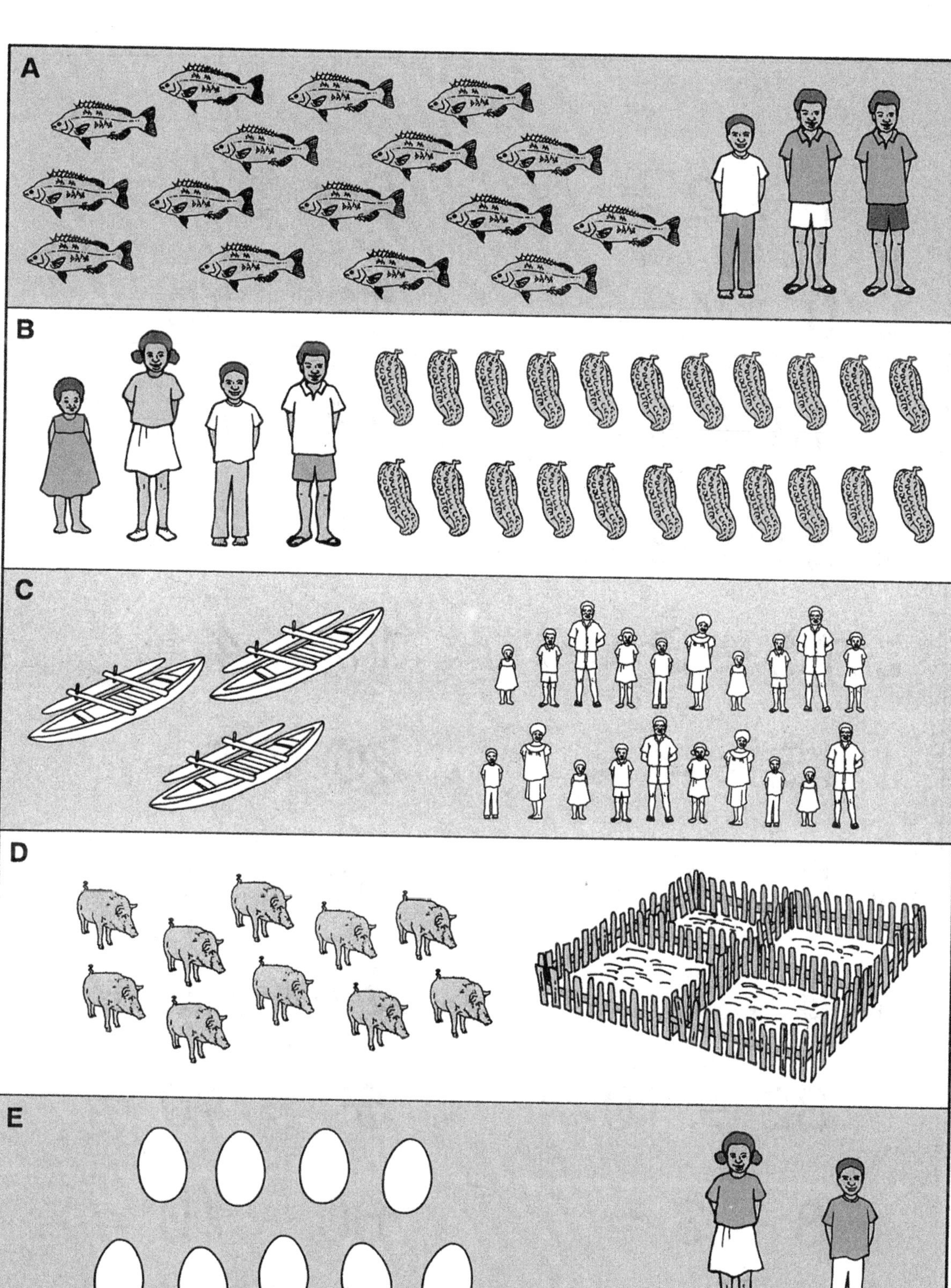
A
B
C
D
E

a $7 \div 2 =$

b $9 \div 4 =$

c $13 \div 5 =$

d $21 \div 2 =$

e $13 \div 3 =$

f $15 \div 4 =$

g $6 \div 4 =$

h $9 \div 2 =$

i $32 \div 10 =$

j $8 \div 3 =$

k $6 \div 5 =$

l $12 \div 10 =$

m $13 \div 4 =$

n $14 \div 4 =$

o $17 \div 4 =$

p $20 \div 3 =$

q $55 \div 10 =$

r $57 \div 10 =$

s $59 \div 10 =$

t $60 \div 10 =$

a $32 \div \square = 8$

b $50 \div \square = 5$

c $\square \div 4 = 5$

d $\square \div 5 = 4$

e $16 \div \square = 2$

f $\square \div 5 = 4$ rem. 1

g $\square \div \square = 2$

h $\square \div \square = 3$ rem. 1

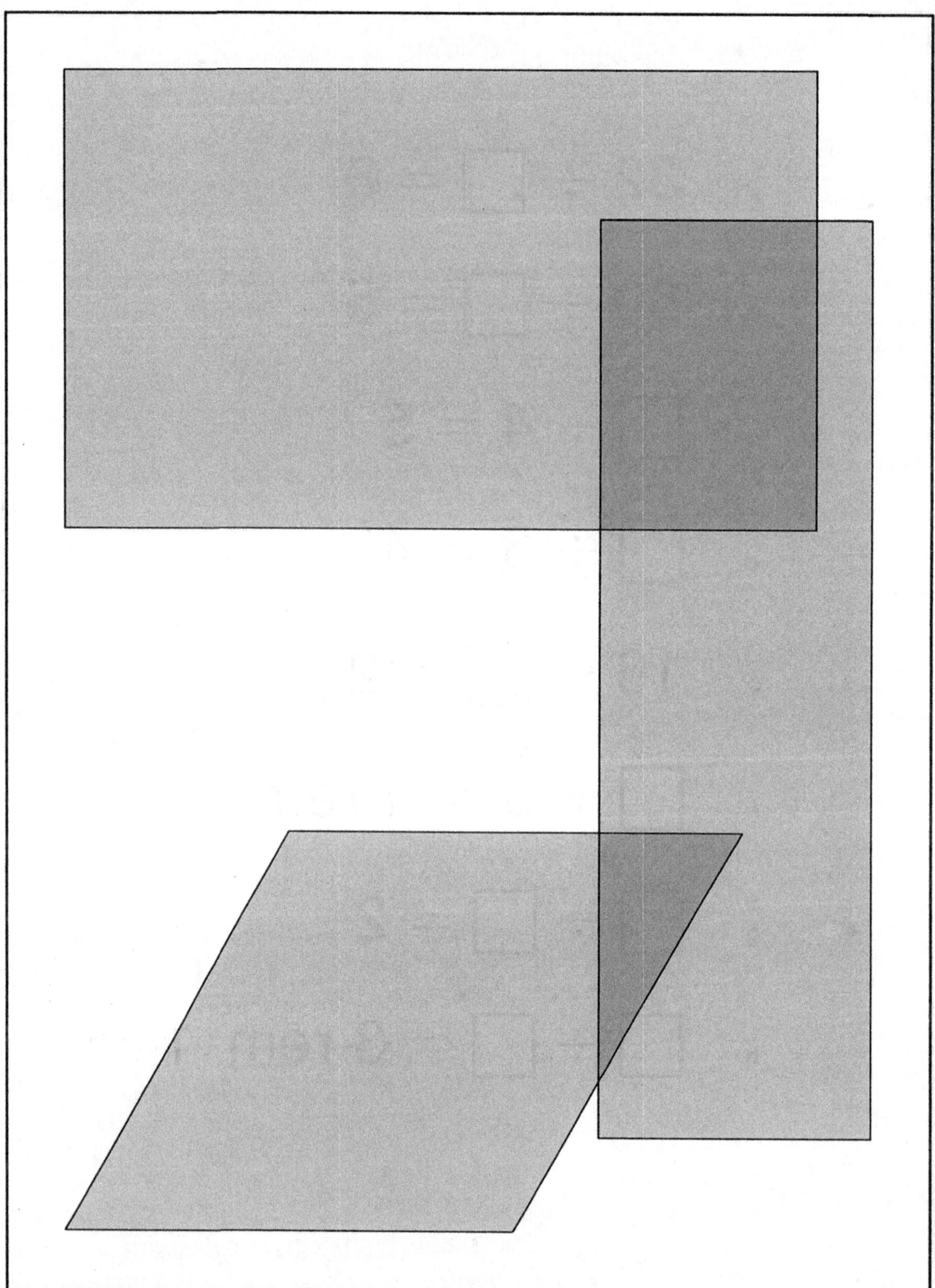

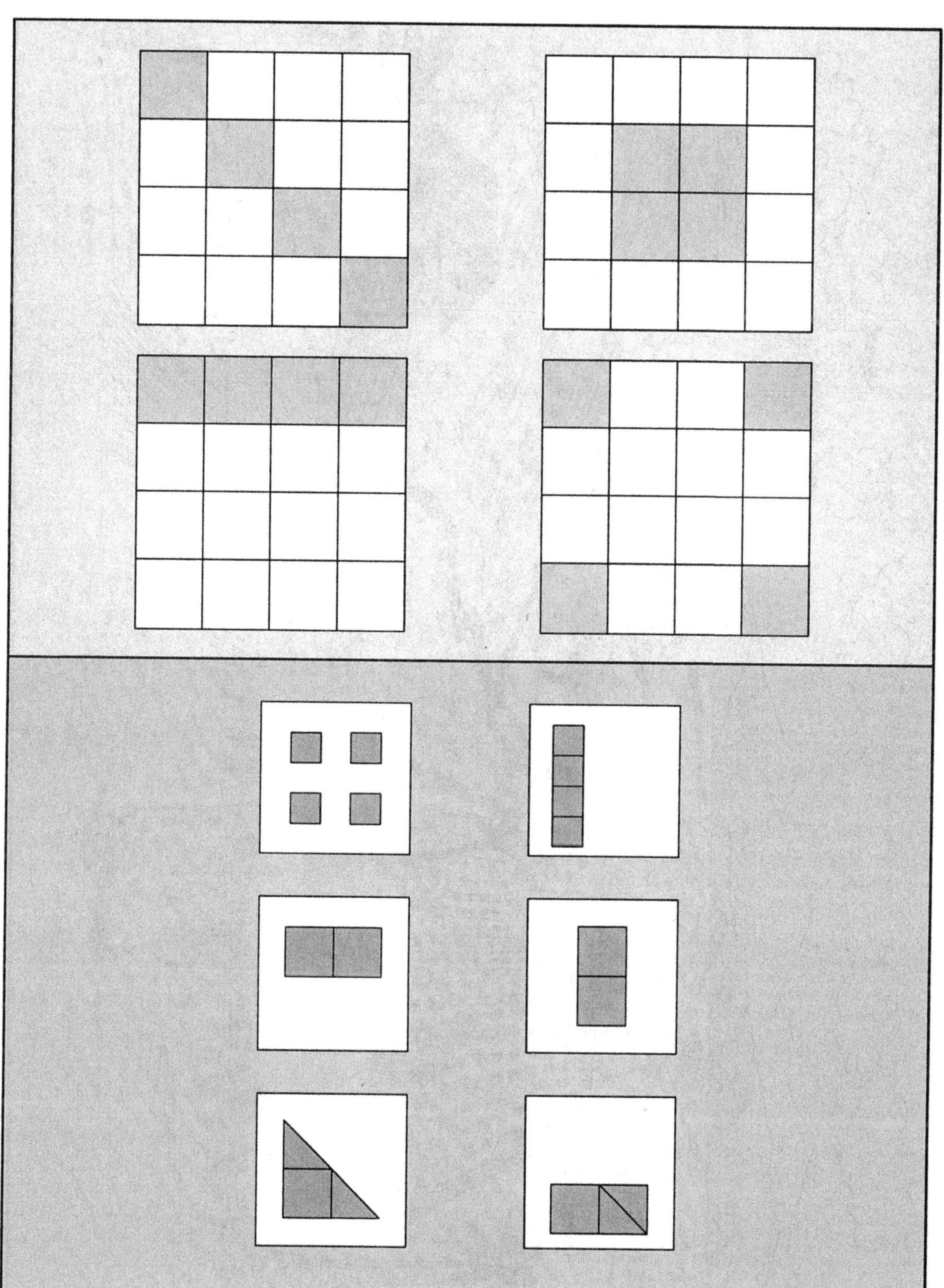

WANTOK
PAPUA NEW GUINEA
Post-Courier
INSIDE TODAY
Court may rule on fees for
Community School
Mathematics 1B

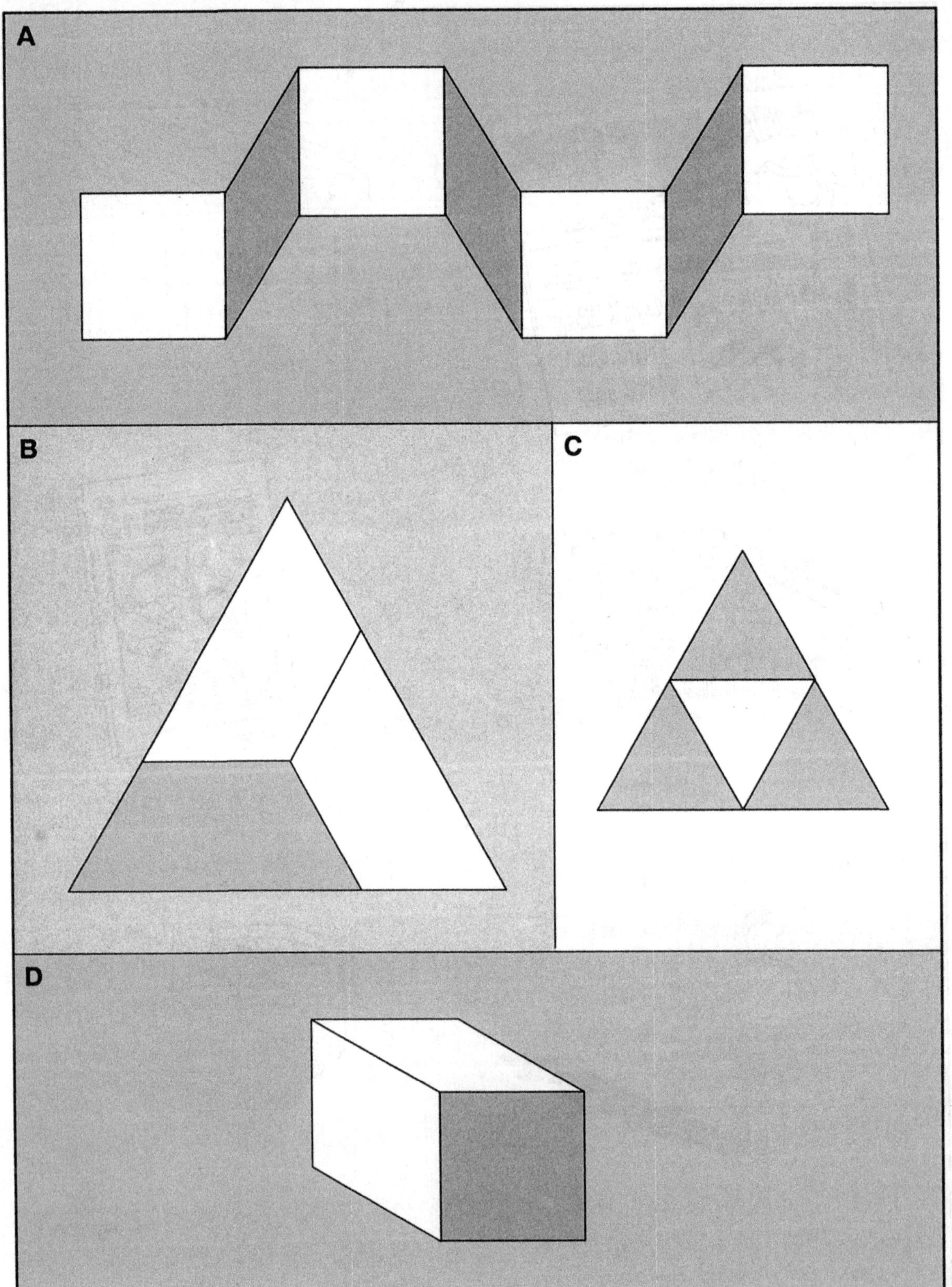
A
B
C
D

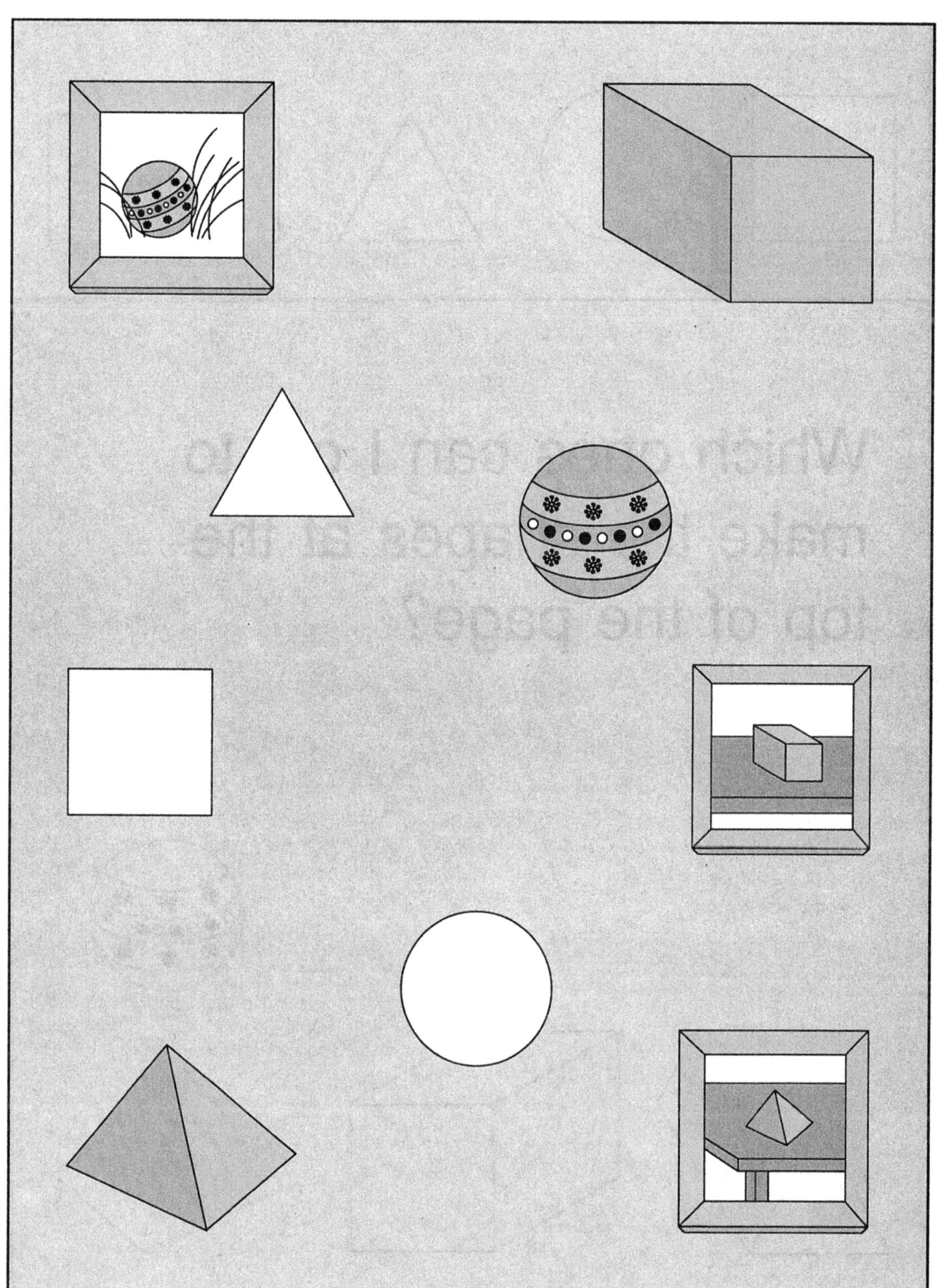

Which ones can I cut to make the shapes at the top of the page?

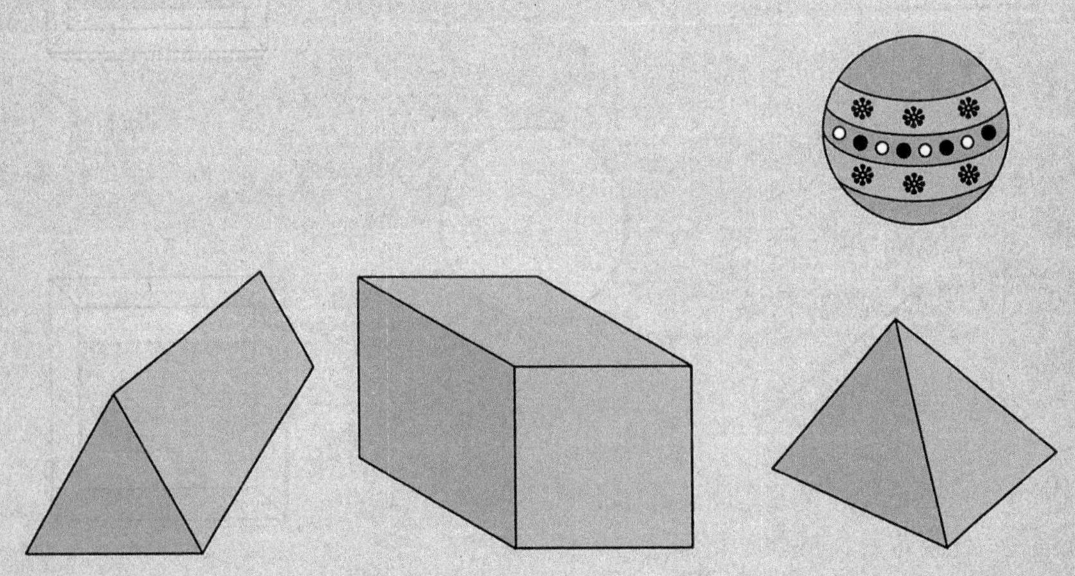

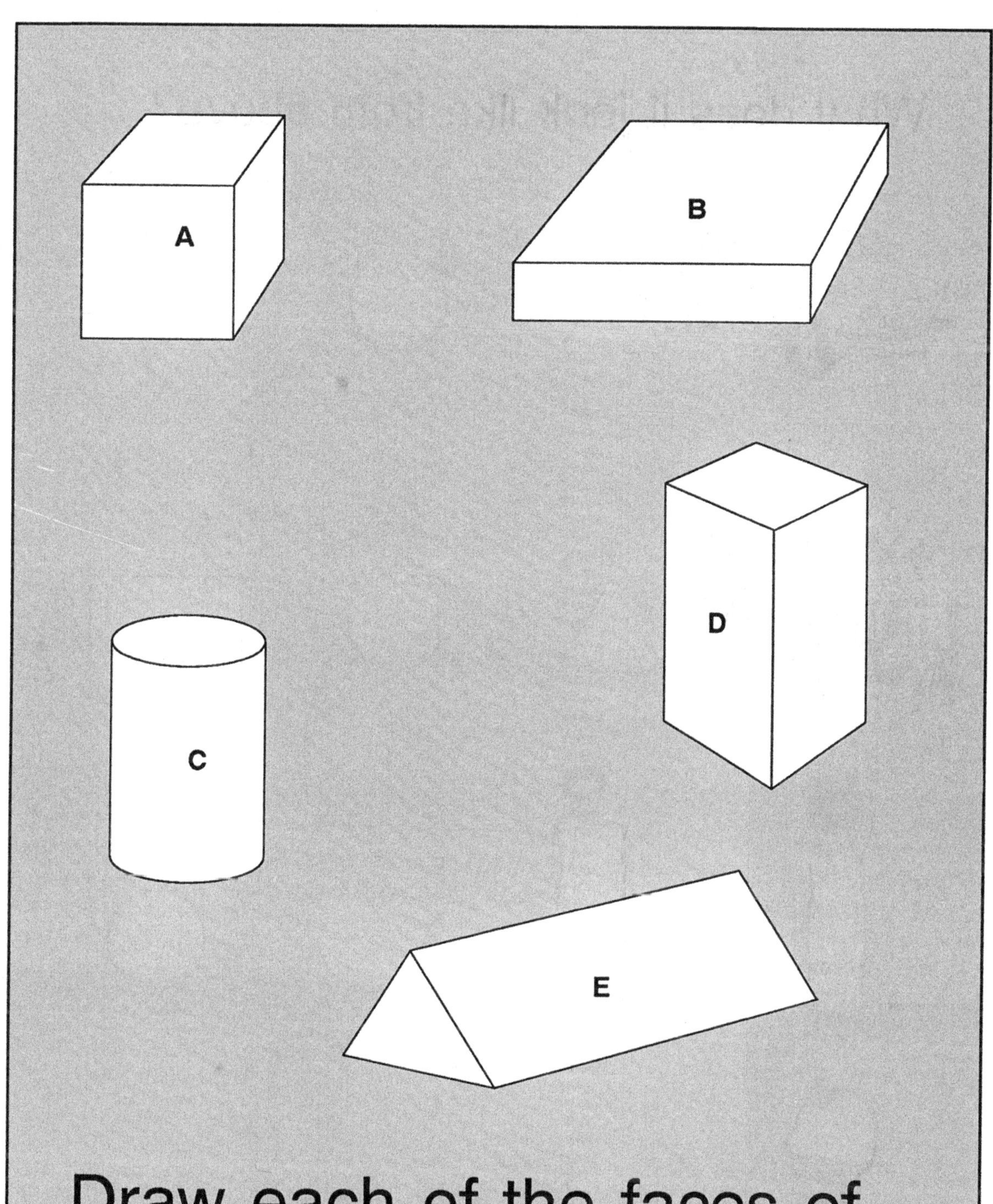

Draw each of the faces of these shapes.

What does it look like from above?

Daniel
Elsie
Mary

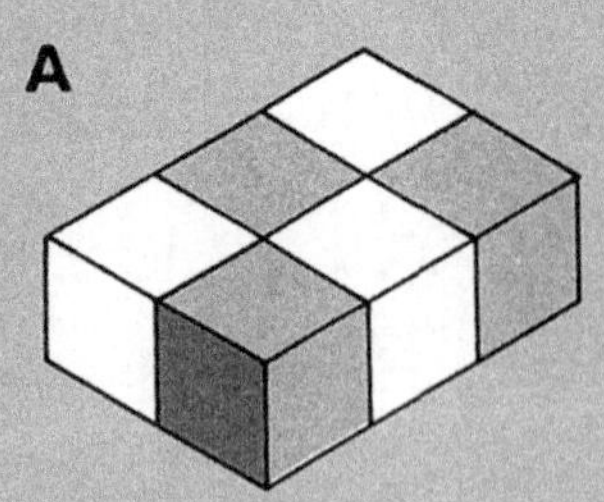

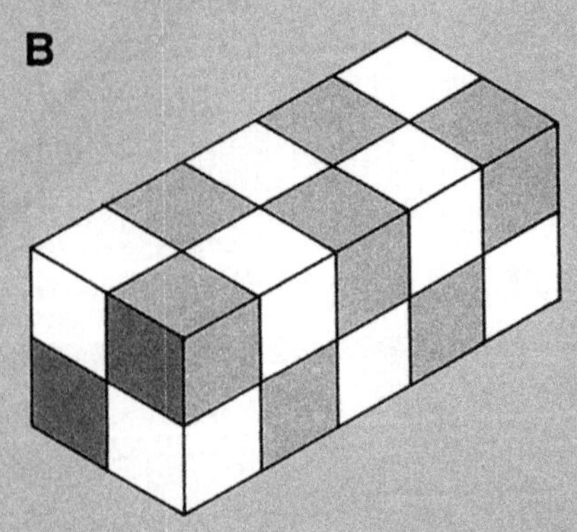

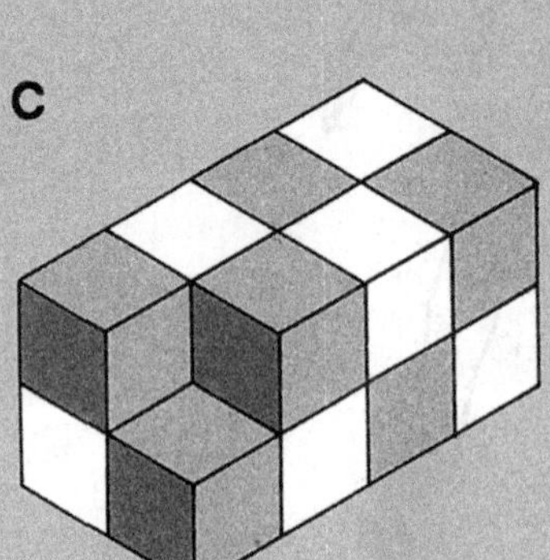

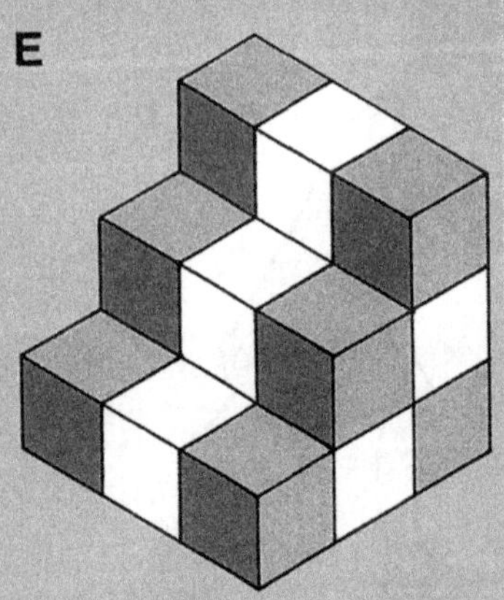

How many cubes are in these shapes?

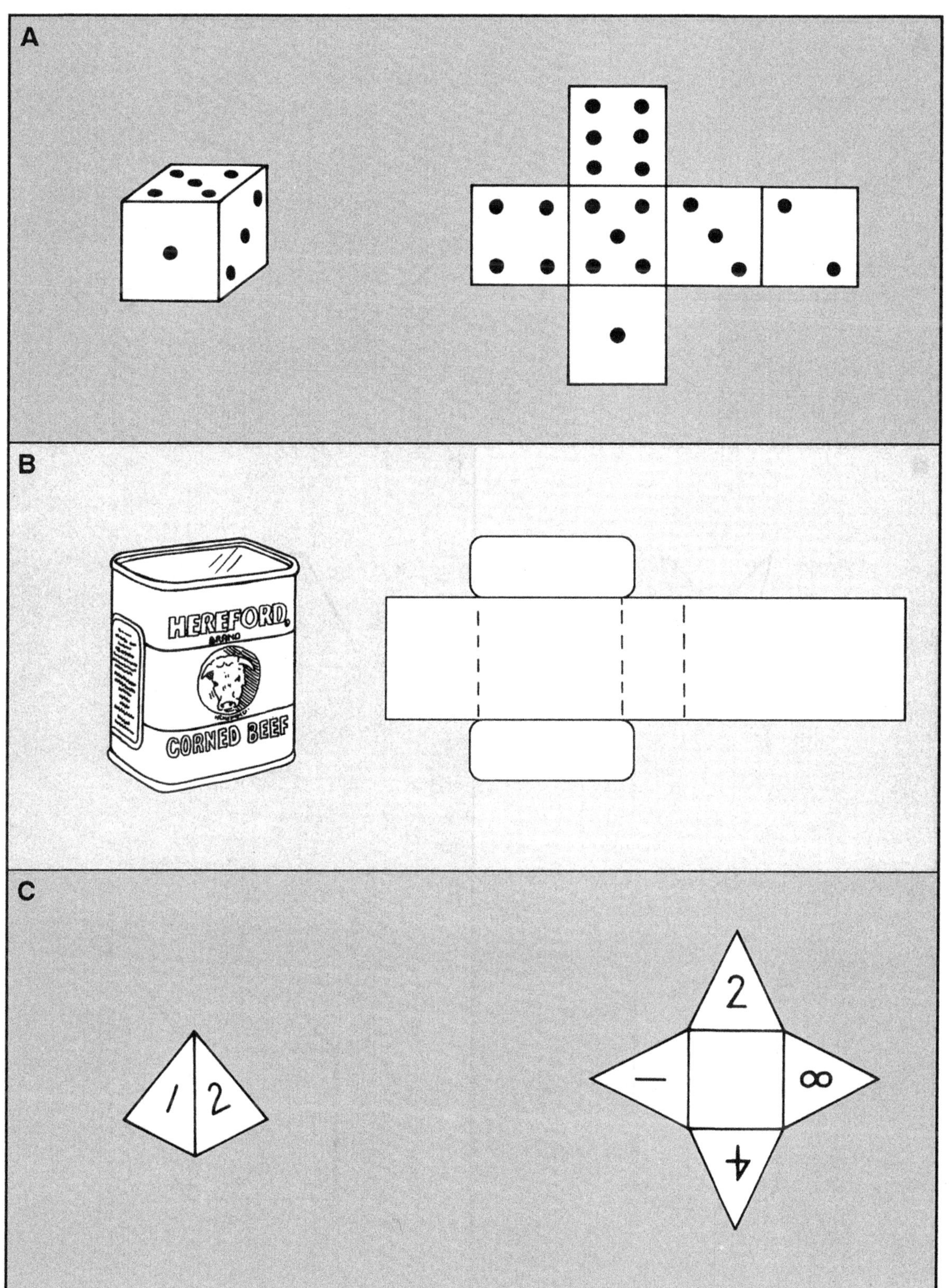
A
B
HEREFORD
CORNED BEEF
C
1 2
2
1
8
4

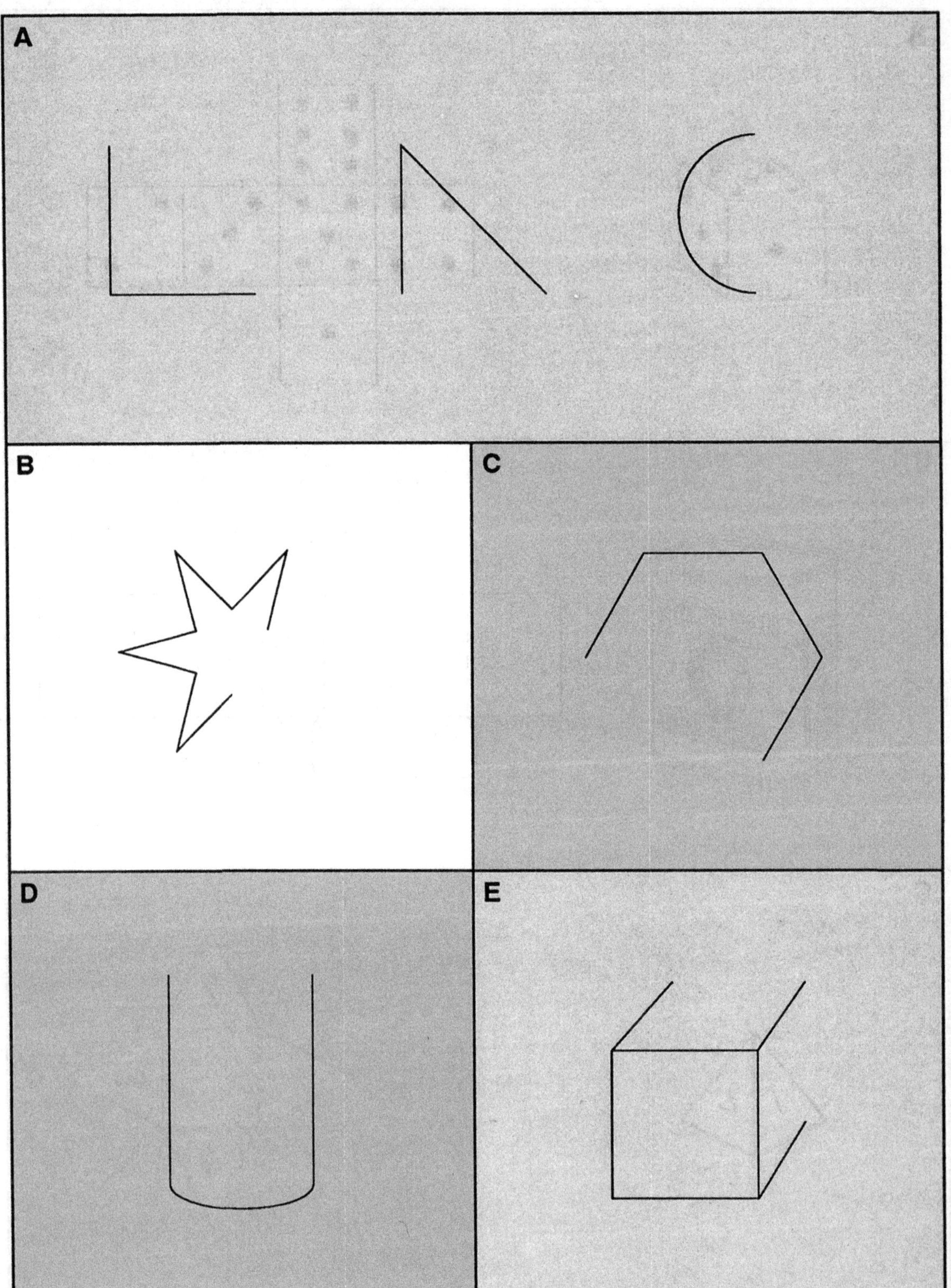
A
B
C
D
E

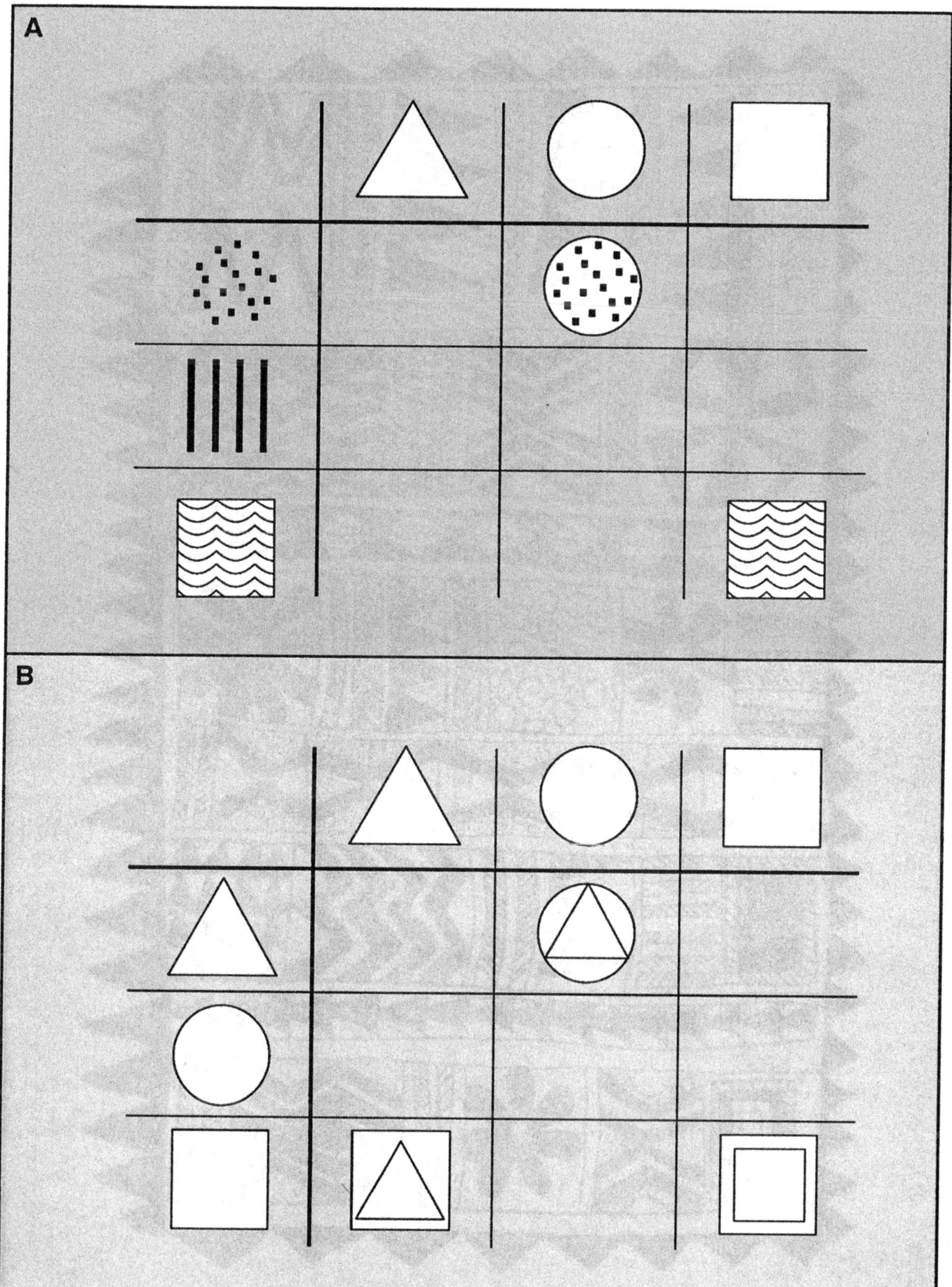
A
B

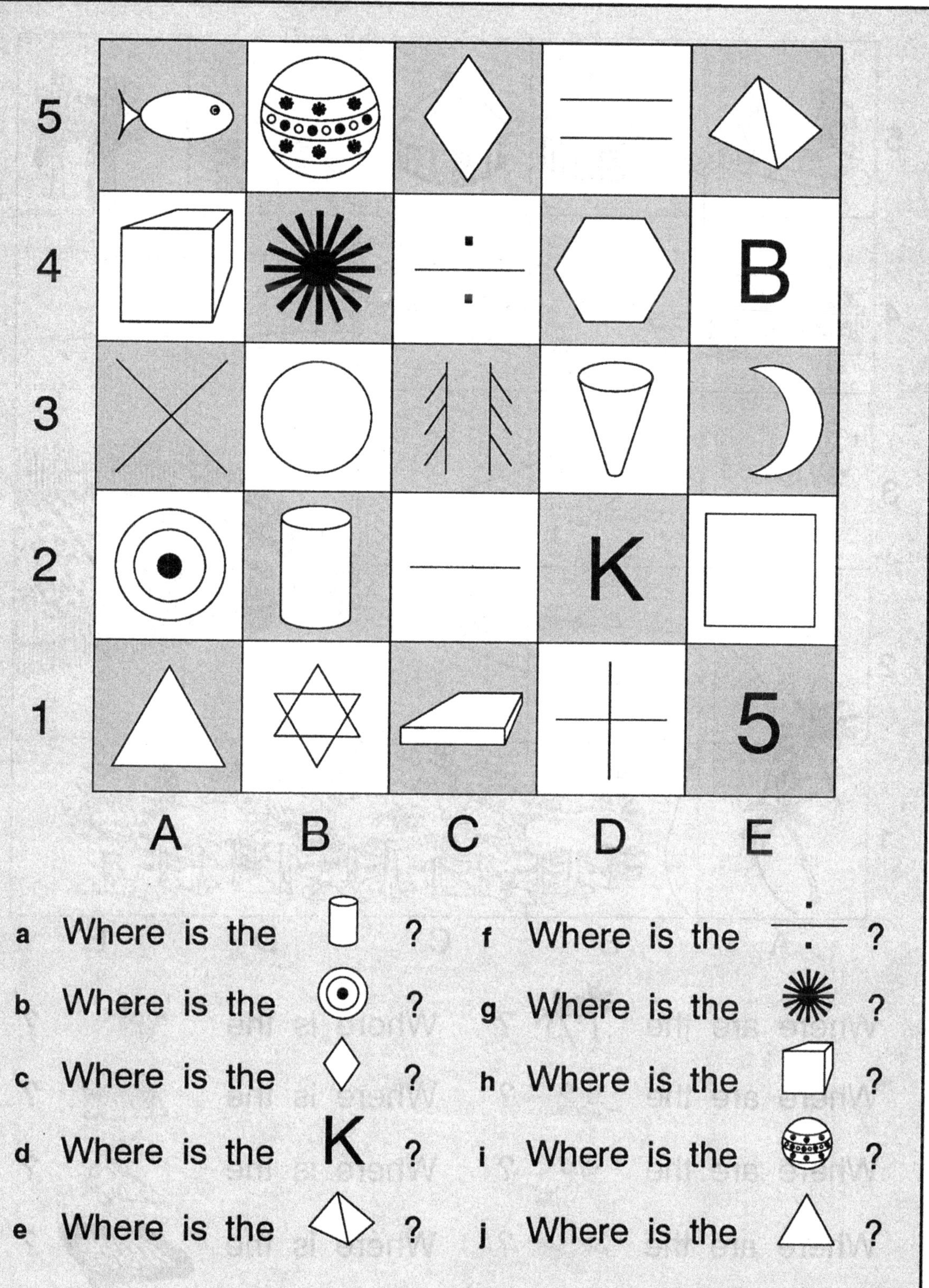

5
4
B
3
2
K
1
5
A
B
C
D
E
a Where is the ?
b Where is the ?
c Where is the ?
d Where is the K ?
e Where is the ?
f Where is the ?
g Where is the ?
h Where is the ?
i Where is the ?
j Where is the ?

5
4
3
2
1
A
B
C
D
E
Where are the ?
Where is the ?
Where are the ?
Where is the ?
Where are the ?
Where is the ?
Where are the ?
Where is the ?

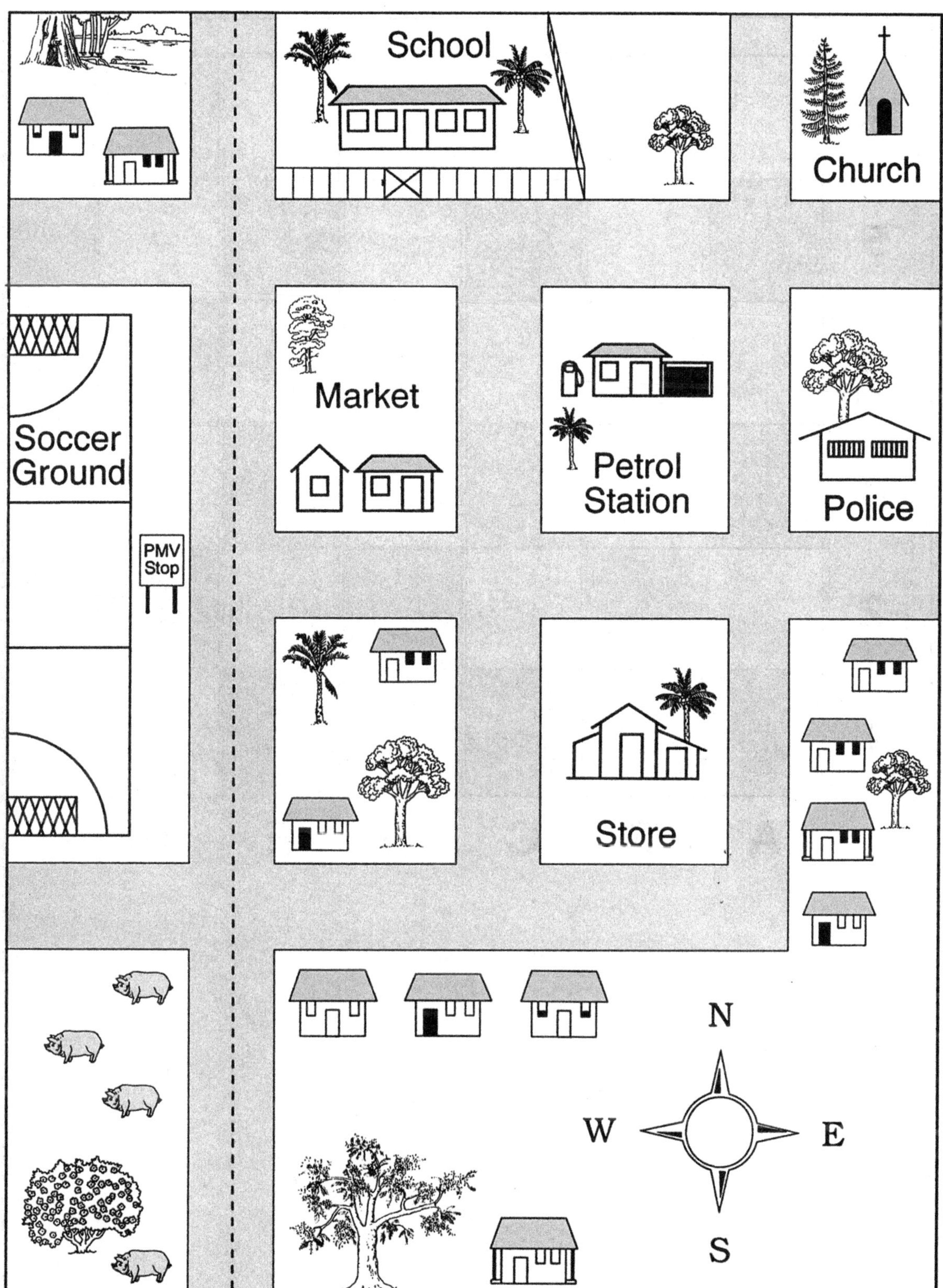
School
Church
Market
Petrol
Station
Police
Soccer
Ground
PMV
Stop
Store
N
W
E
S

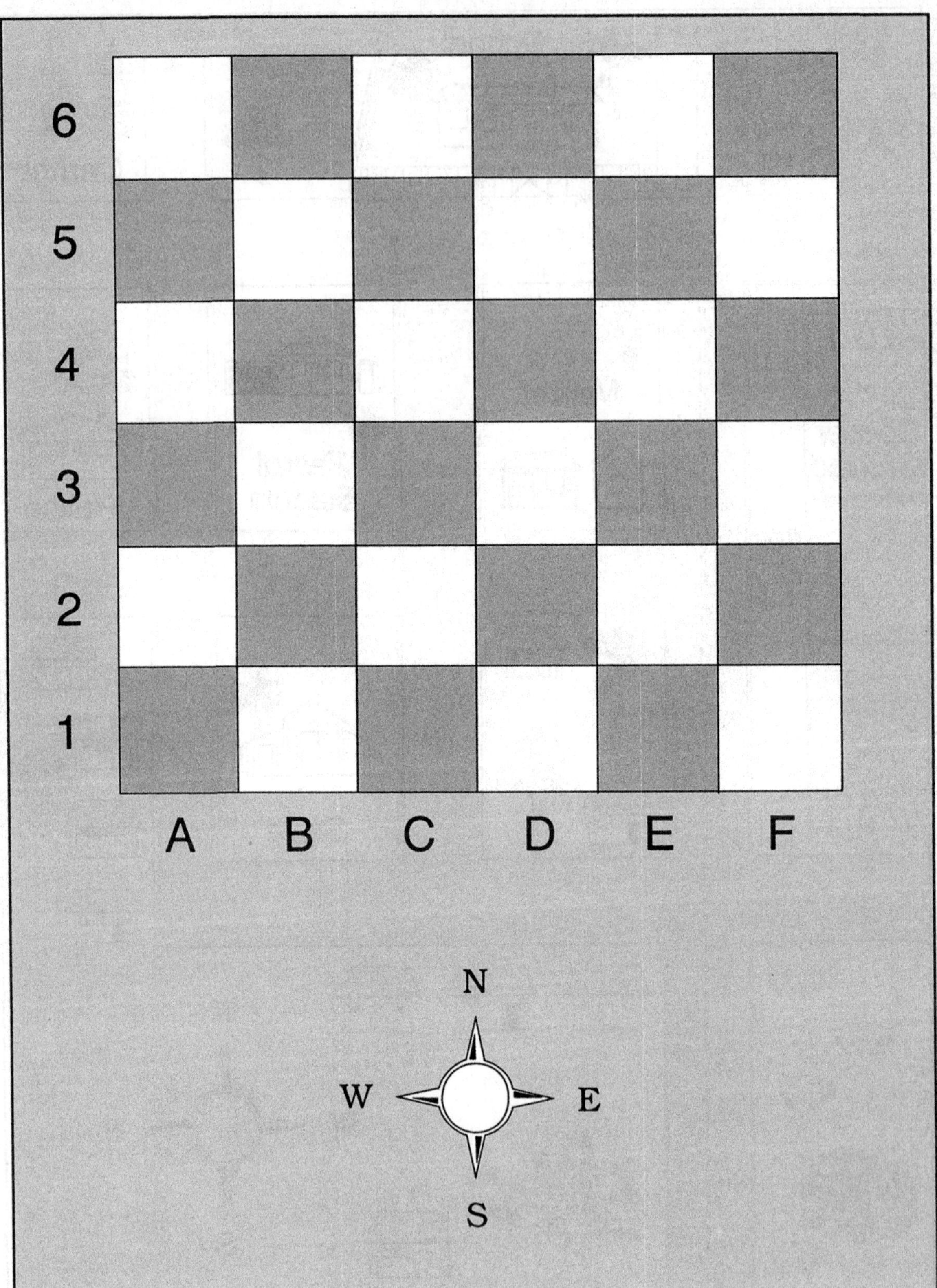
6
5
4
3
2
1
A
B
C
D
E
F
N
W
E
S

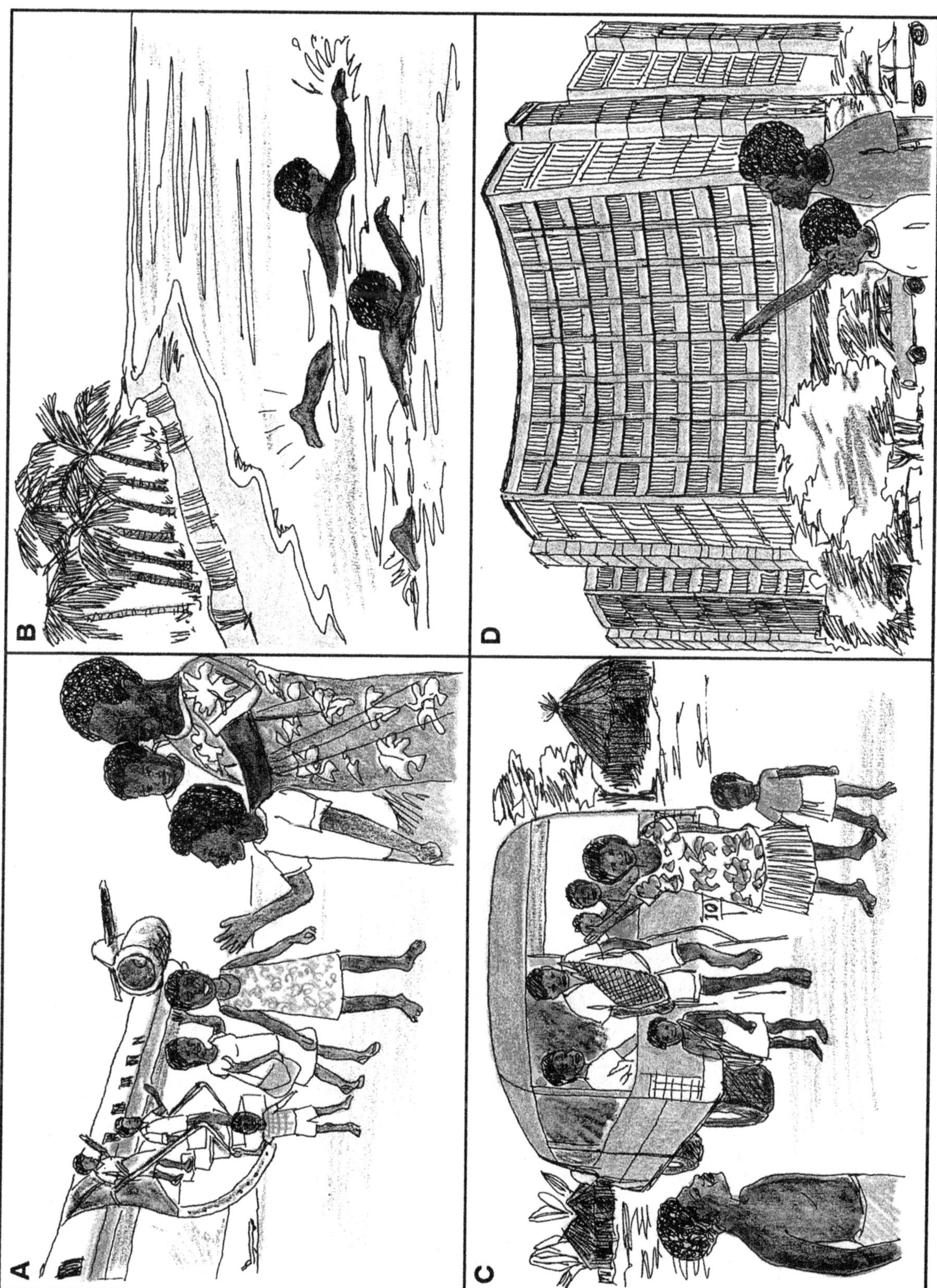
A
B
C
D

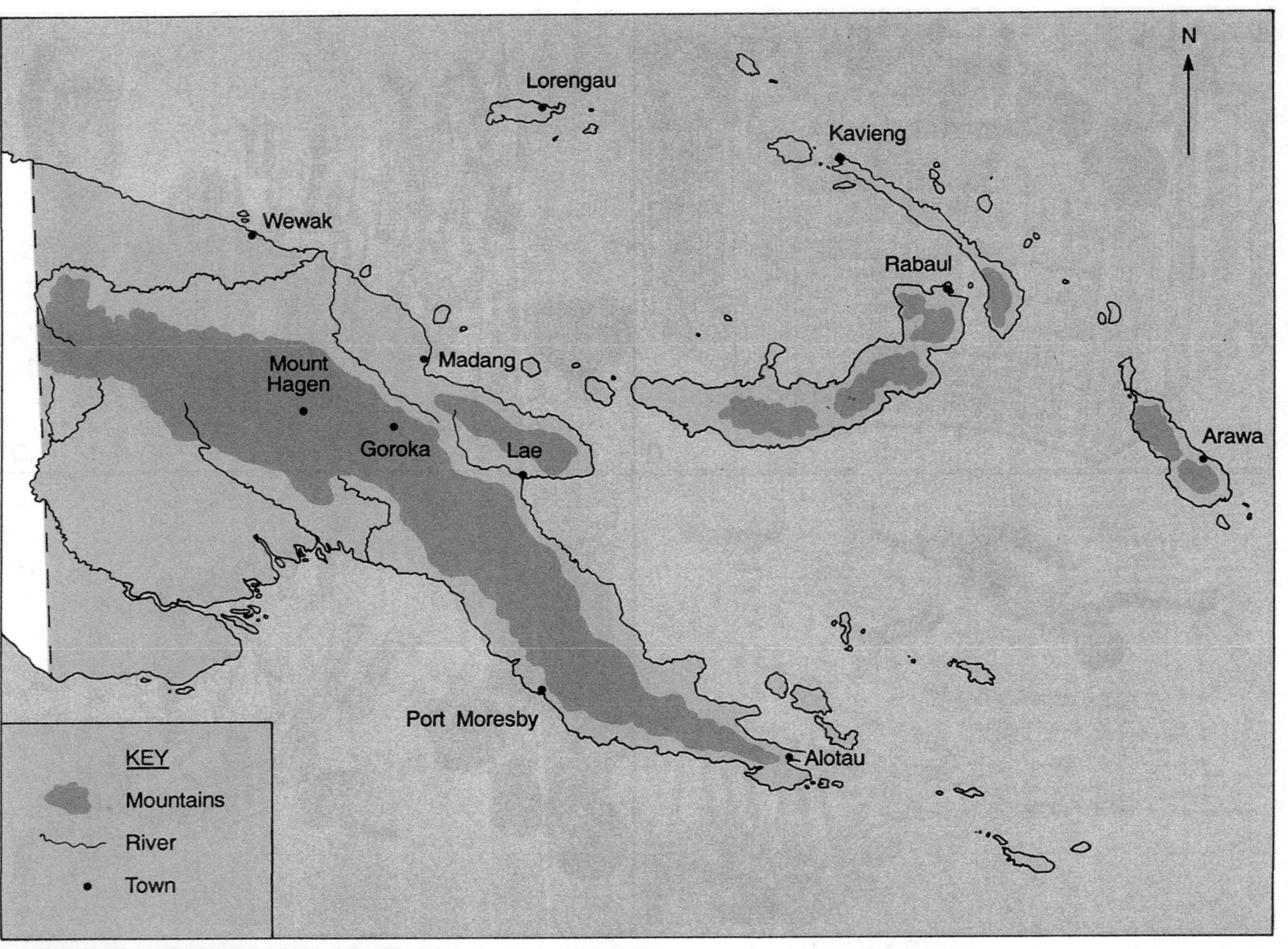
N
Lorengau
Kavieng
Wewak
Rabaul
Madang
Mount
Hagen
Goroka
Lae
Arawa
Port Moresby
Alotau
KEY
Mountains
River
Town

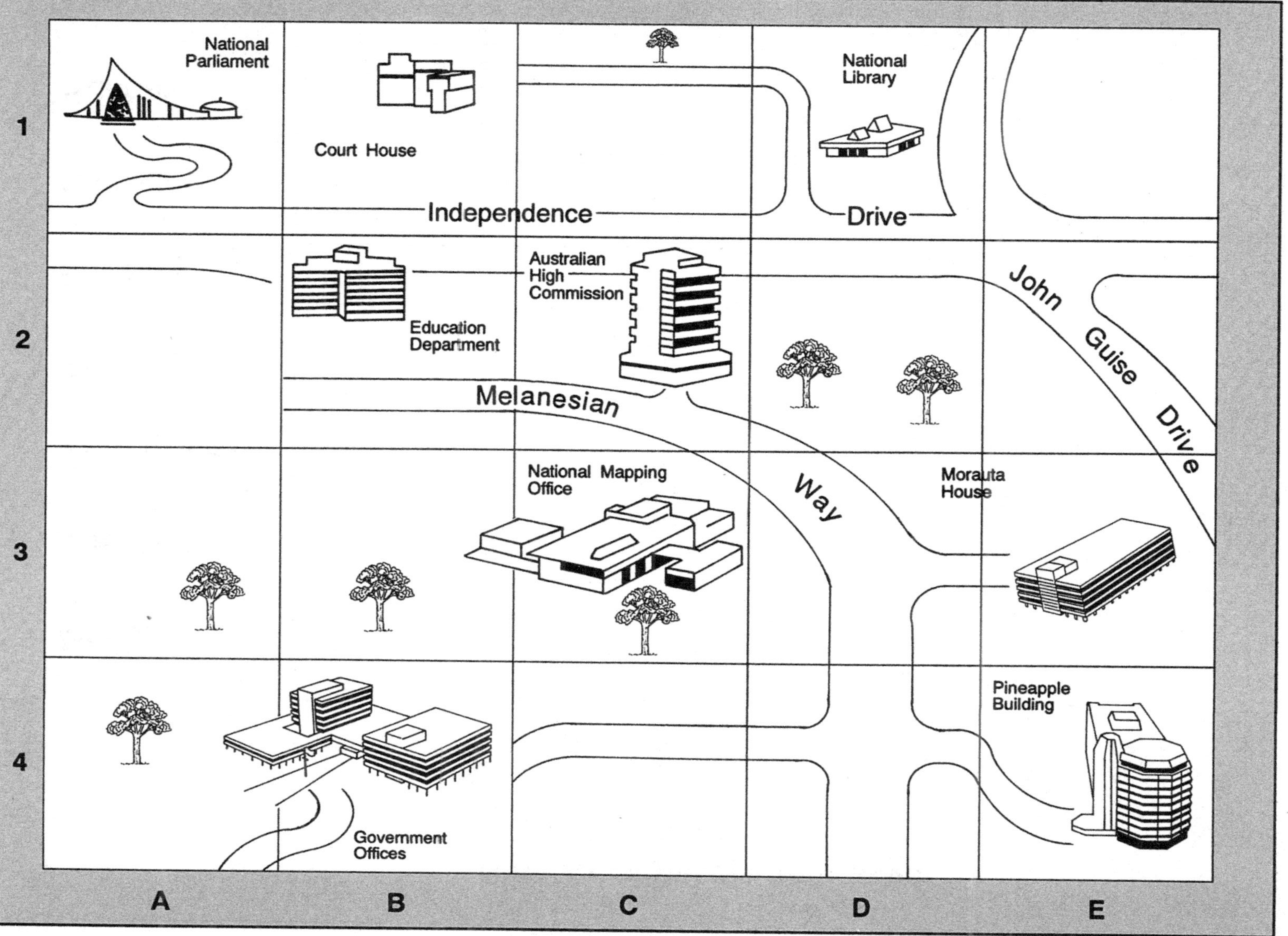
National Parliament
Court House
Independence Drive
National Library
John Guise Drive
Education Department
Australian High Commission
Melanesian Way
National Mapping Office
Morauta House
Pineapple Building
Government Offices
1
2
3
4
A
B
C
D
E

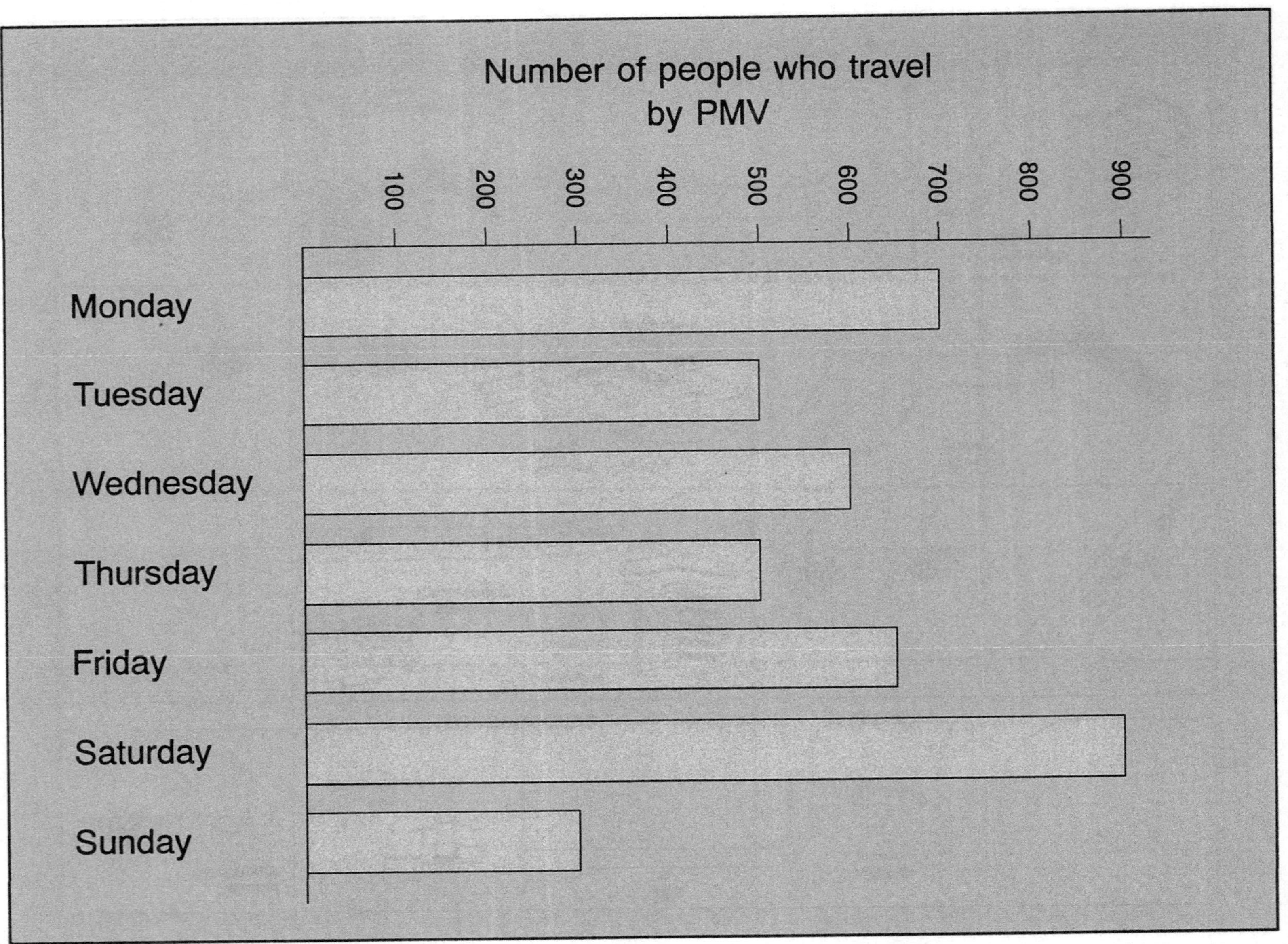

Number of people who travel
by PMV
100
200
300
400
500
600
700
800
900
Monday
Tuesday
Wednesday
Thursday
Friday
Saturday
Sunday

40kg
1kg
10kg
20kg
15kg
Coke

TALAIR
DASH 8
TALAIR
12kg
12kg
12kg
12kg
16kg
16kg
25kg
15kg
15kg
15kg
21kg
21kg

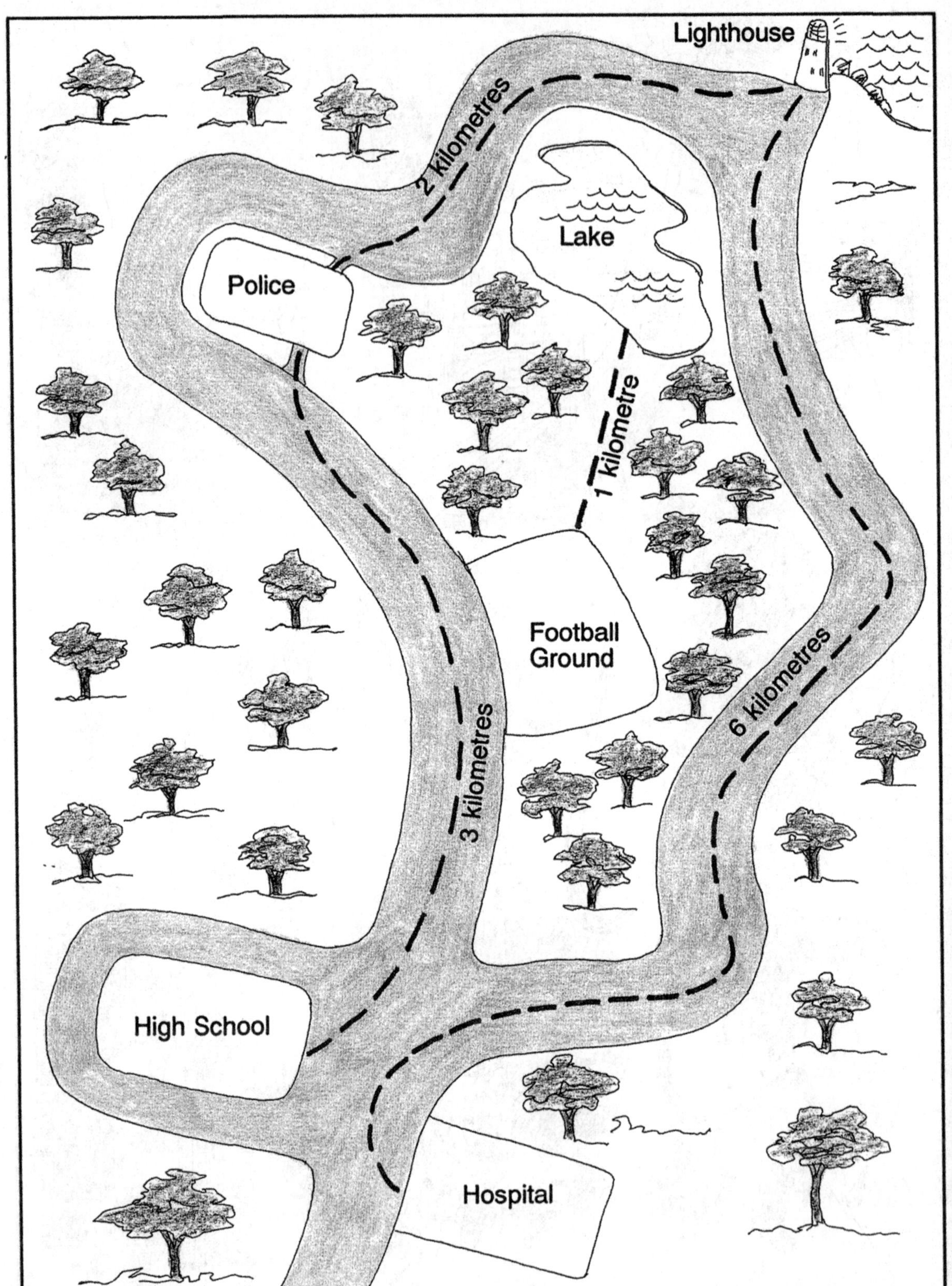
Lighthouse
2 kilometres
Lake
Police
1 kilometre
Football
Ground
3 kilometres
6 kilometres
High School
Hospital

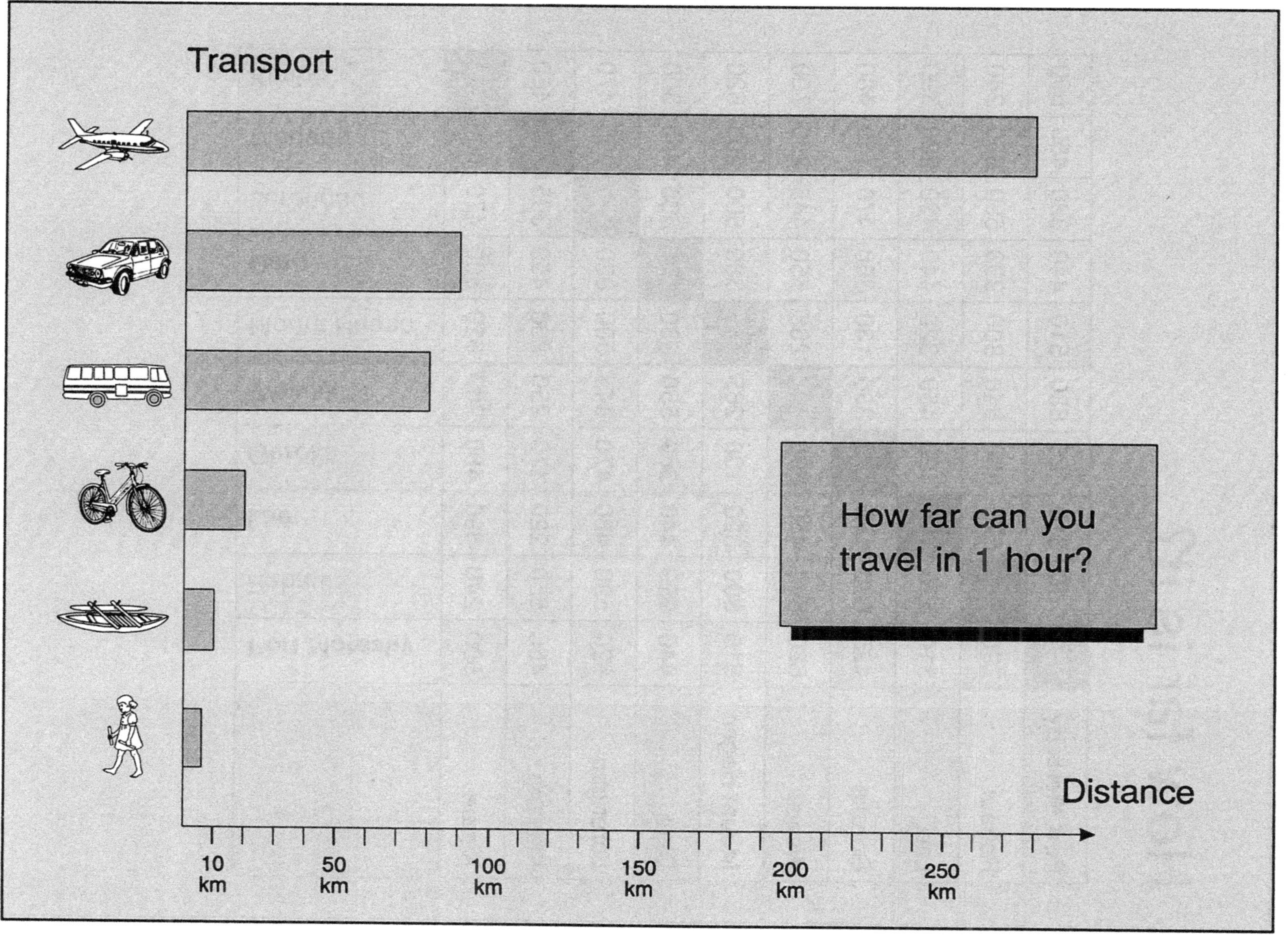
Transport
How far can you travel in 1 hour?
Distance
10 km
50 km
100 km
150 km
200 km
250 km

How far is it?

	Port Moresby	Rabaul	Lae	Goroka	Wewak	Mount Hagen	Daru	Lorengau	Madang	Kimbe
Port Moresby		790	310	425	680	510	440	820	450	500
Rabaul	790		570	680	840	800	999	590	650	240
Lae	310	570		190	450	320	440	480	220	390
Goroka	425	680	190		330	130	360	470	110	480
Wewak	680	840	450	330		265	550	445	250	700
Mount Hagen	510	800	320	130	265		380	500	200	620
Daru	440	999	440	360	550	380		830	450	800
Lorengau	820	590	480	470	445	500	830		395	470
Madang	450	650	220	110	250	200	450	395		460
Kimbe	500	240	390	480	700	620	800	470	460	

MENU

Bunch of peanuts 30t

Coke/Fanta 80t

Apple 25t

Long life milk 70t

Peanut butter K2.19

Boiled egg 30t

Bunch of 3 bananas 60t

Tin fish 90t

No.1 pineapple juice 60t

Loaf of bread 70t

Tin of Milo K1.69

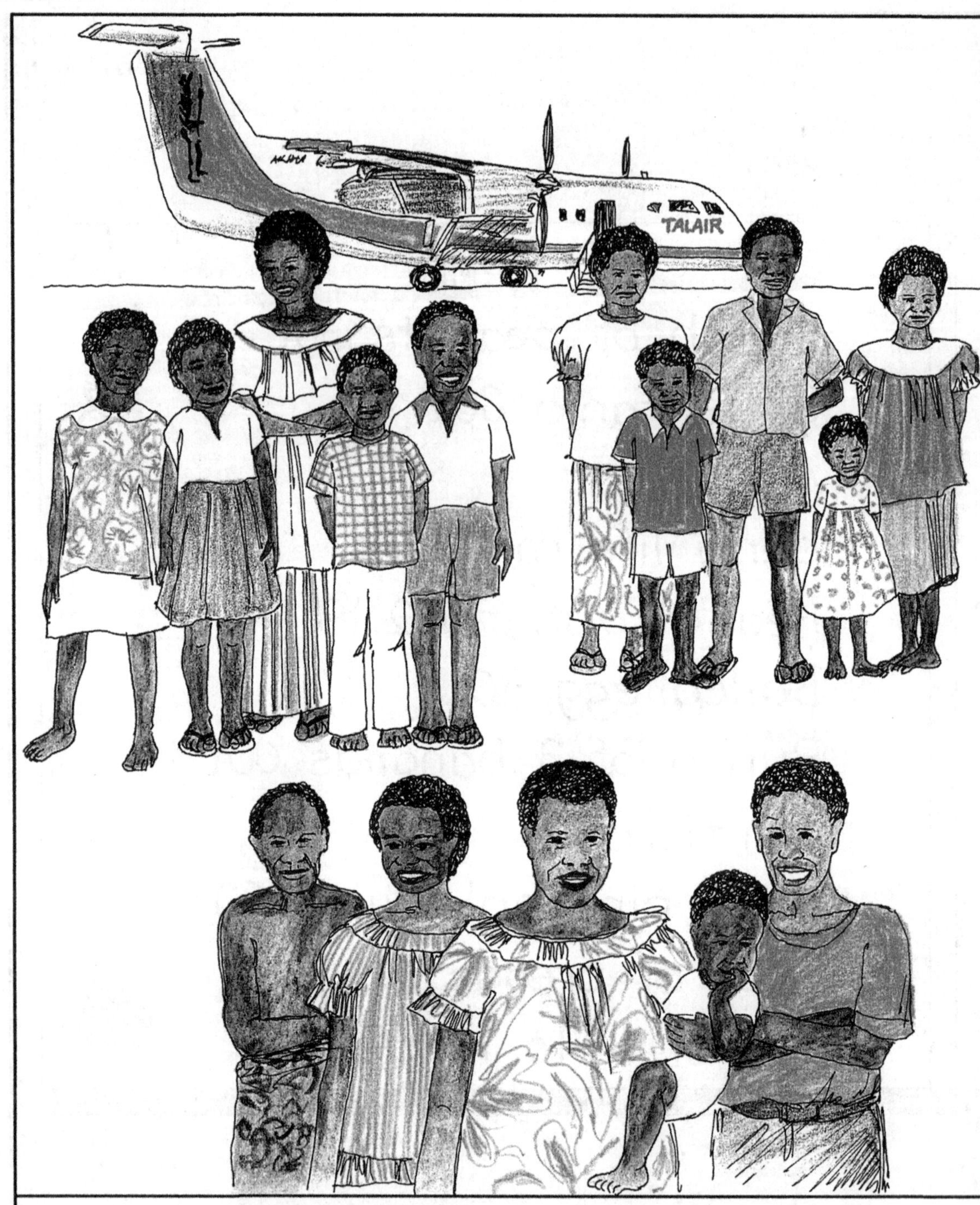

Which family will pay the most?

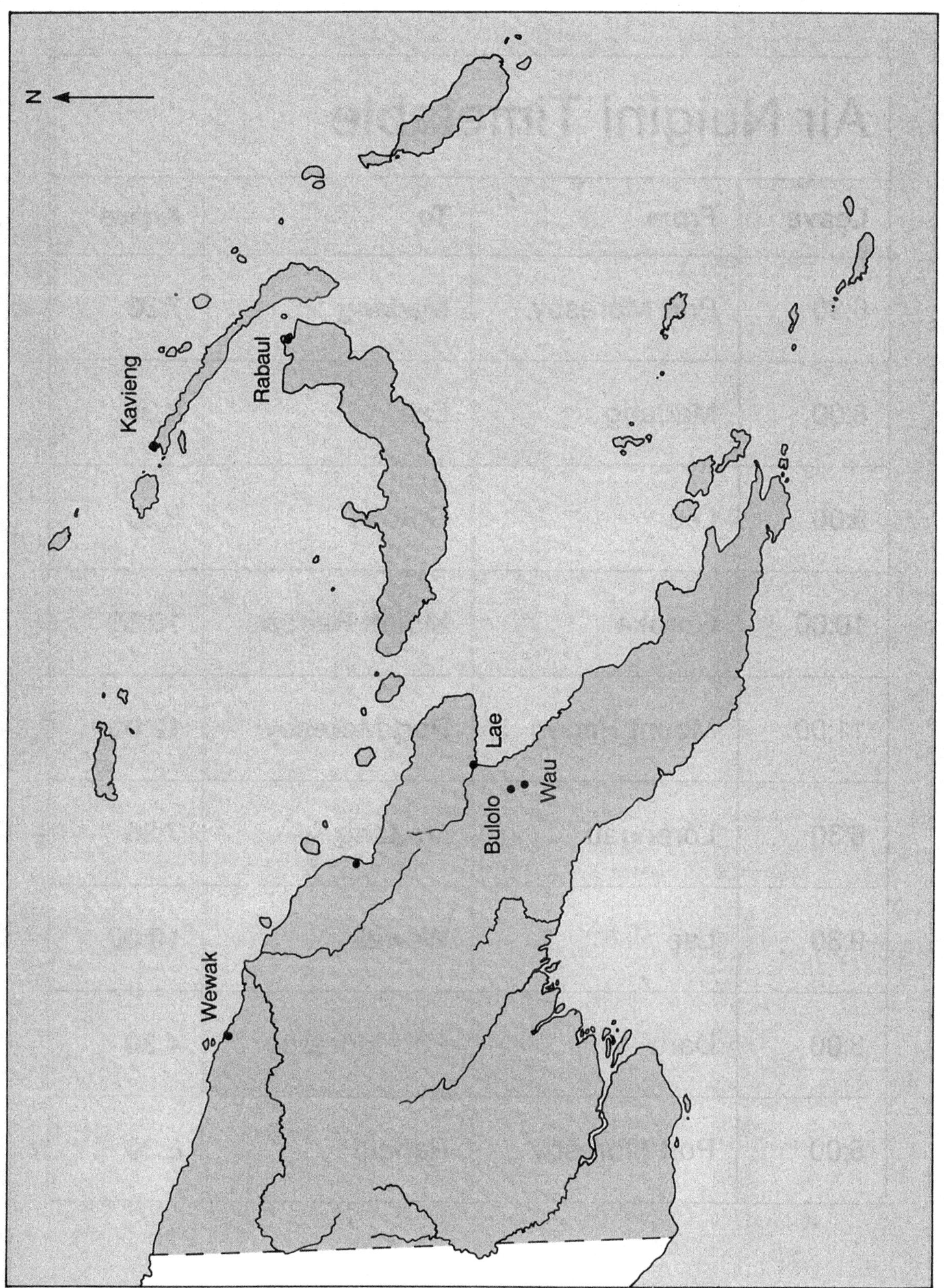
N
Kavieng
Rabaul
Lae
Bulolo
Wau
Wewak

Air Nuigini Timetable

Leave	From	To	Arrive
6:30	Port Moresby	Madang	7:30
8:00	Madang	Lae	8:30
9:00	Lae	Goroka	9:30
10:00	Goroka	Mount Hagen	10:30
11:00	Mount Hagen	Port Moresby	12:00
6:30	Lorengau	Madang	7:30
9:30	Lae	Wewak	10:00
3:00	Daru	Port Moresby	4:30
5:00	Port Moresby	Rabaul	6:30

Douglas DC-3/Dakota

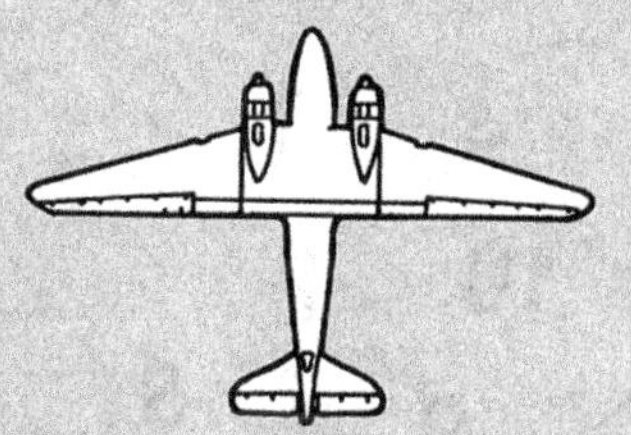

Span: 29m
Length: 20m
Passengers: 36
Range: 2,430 km

Cessna 401/402

Span: 12m
Length: 11m
Passengers: 9
Range: 2,637 km

Fokker F28 Fellowship

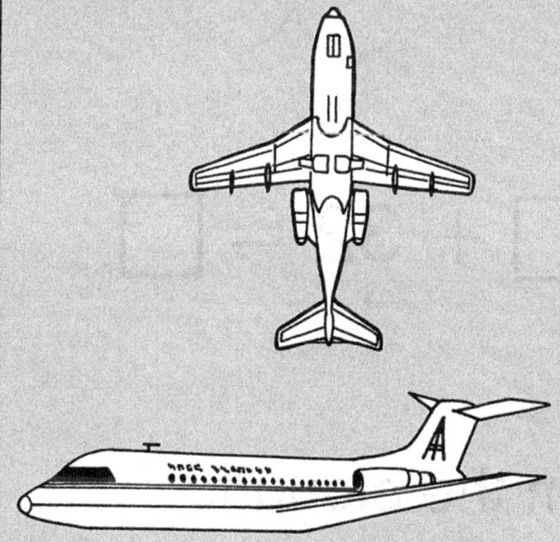

Span: 25m
Length: 21m
Passengers: 65

Britten-Norman Islander

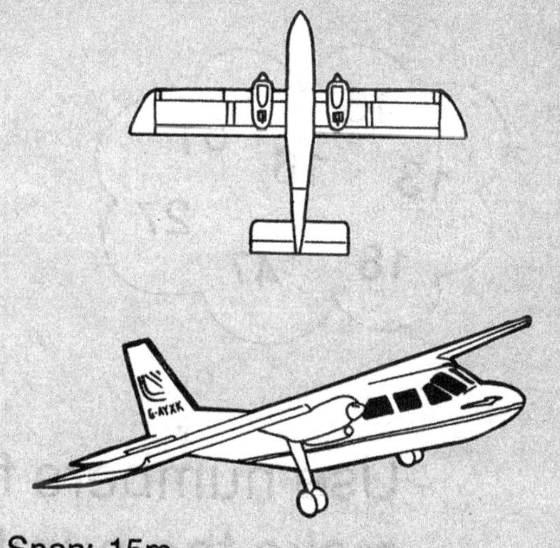

Span: 15m
Length: 12m
Passengers: 10
Range: 1,400 km

Use numbers from the cloud to make the number sentences correct.

A = 26	B = 25	C = 24
D = 23	E = 22	F = 21
G = 20	H = 19	I = 18
J = 17	K = 16	L = 15
M = 14	N = 13	O = 12
P = 11	Q = 10	R = 9
S = 8	T = 7	U = 6
V = 5	W = 4	X = 3
Y = 2	Z = 1	

What is your name worth?

January

S	M	T	W	T	F	S
1	2	3	4	5	6	7
8	9	10	11	12	13	14
15	16	17	18	19	20	21
22	23	24	25	26	27	28
29	30	31				

February

S	M	T	W	T	F	S
			1	2	3	4
5	6	7	8	9	10	11
12	13	14	15	16	17	18
19	20	21	22	23	24	25
26	27	28				

March

S	M	T	W	T	F	S
			1	2	3	4
5	6	7	8	9	10	11
12	13	14	15	16	17	18
19	20	21	22	23	24	25
26	27	28	29	30	31	

April

S	M	T	W	T	F	S
30						1
2	3	4	5	6	7	8
9	10	11	12	13	14	15
16	17	18	19	20	21	22
23	24	25	26	27	28	29

May

S	M	T	W	T	F	S
	1	2	3	4	5	6
7	8	9	10	11	12	13
14	15	16	17	18	19	20
21	22	23	24	25	26	27
28	29	30	31			

June

S	M	T	W	T	F	S
				1	2	3
4	5	6	7	8	9	10
11	12	13	14	15	16	17
18	19	20	21	22	23	24
25	26	27	28	29	30	

July

S	M	T	W	T	F	S
30	31					1
2	3	4	5	6	7	8
9	10	11	12	13	14	15
16	17	18	19	20	21	22
23	24	25	26	27	28	29

August

S	M	T	W	T	F	S
		1	2	3	4	5
6	7	8	9	10	11	12
13	14	15	16	17	18	19
20	21	22	23	24	25	26
27	28	29	30	31		

September

S	M	T	W	T	F	S
					1	2
3	4	5	6	7	8	9
10	11	12	13	14	15	16
17	18	19	20	21	22	23
24	25	26	27	28	29	30

October

S	M	T	W	T	F	S
1	2	3	4	5	6	7
8	9	10	11	12	13	14
15	16	17	18	19	20	21
22	23	24	25	26	27	28
29	30	31				

November

S	M	T	W	T	F	S
			1	2	3	4
5	6	7	8	9	10	11
12	13	14	15	16	17	18
19	20	21	22	23	24	25
26	27	28	29	30		

December

S	M	T	W	T	F	S
31					1	2
3	4	5	6	7	8	9
10	11	12	13	14	15	16
17	18	19	20	21	22	23
24	25	26	27	28	29	30

1	2	3	4	5	6	7	8	9	10
11	12	13	14	15	16	17	18	19	20
21	22	23	24	25	26	27	28	29	30
31	32	33	34	35	36	37	38	39	40
41	42	43	44	45	46	47	48	49	50
51	52	53	54	55	56	57	58	59	60
61	62	63	64	65	66	67	68	69	70
71	72	73	74	75	76	77	78	79	80
81	82	83	84	85	86	87	88	89	90
91	92	93	94	95	96	97	98	99	100

6 4 3 2

Use any 3 of these numbers to finish these number sentences.

a $\square - \square - \square = 0$

b $\square + \square - \square = 1$

c $\square + \square + \square = 9$

d $\square + \square - \square = 8$

e $\square + \square - \square = 7$

f $\square - \square + \square = 5$

g $\square + \square - \square = 4$

h $\square + \square - \square = 3$

Use numbers from the cloud to make the number sentences correct.

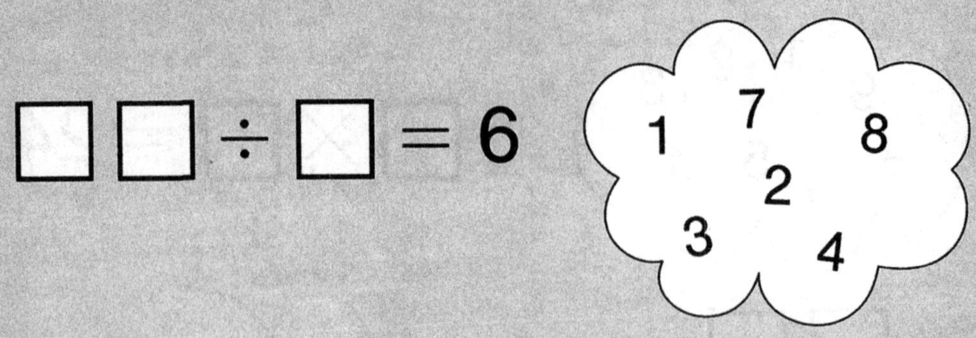

3 2 8 4 5

□□ ÷ □ = □

□□ ÷ □ = □

2 1 7 8 3

Use numbers from the cloud to make the number sentences correct.

A

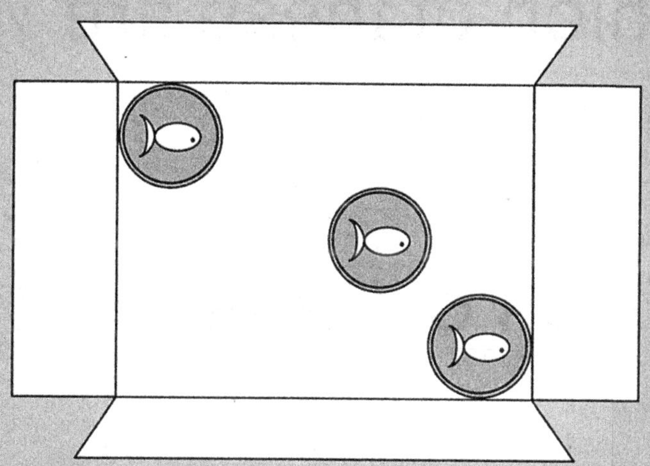

How many tins fill the box?

B

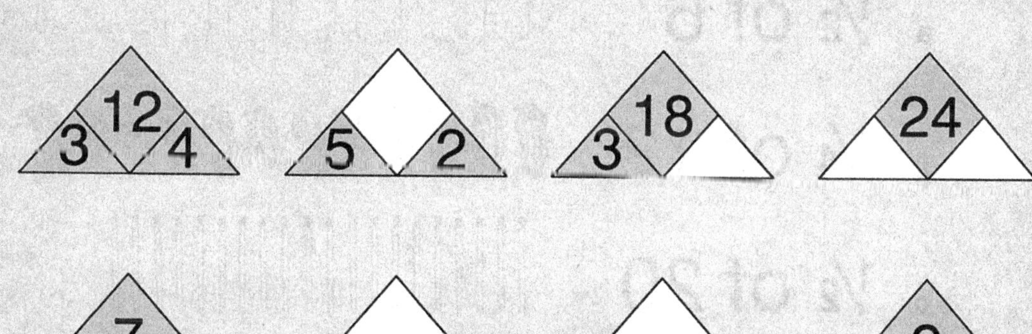

Work out the missing numbers.

1 Which of these are ½?

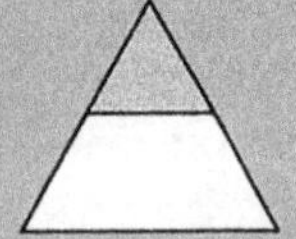
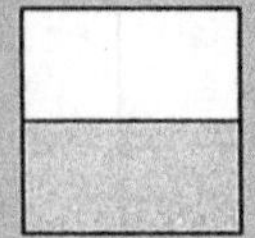
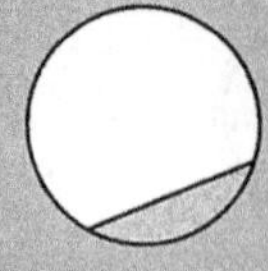
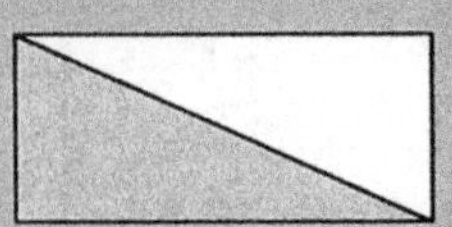

2 Which of these are ¼?

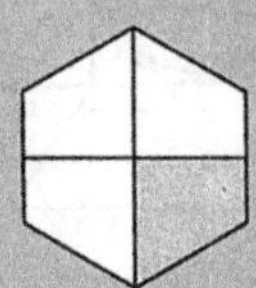
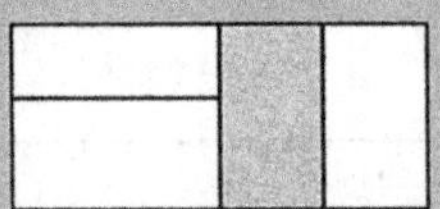
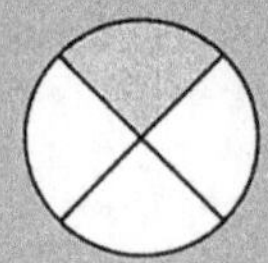

3 Find

a ½ of 6

b ¼ of 12

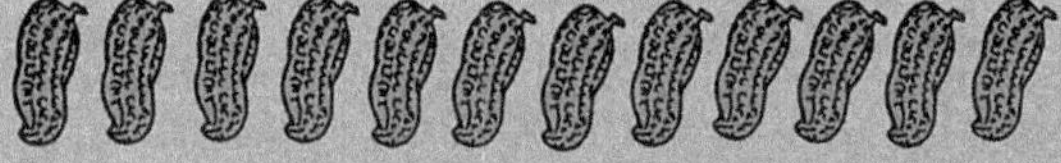

c ½ of 20

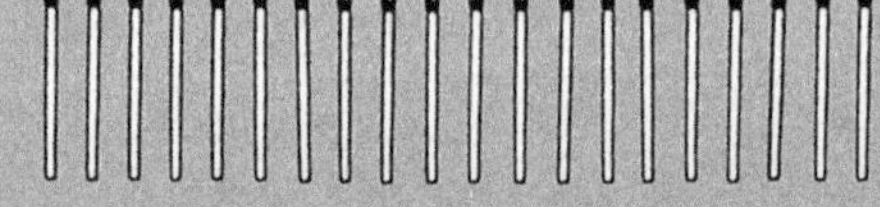

d ¼ of 20

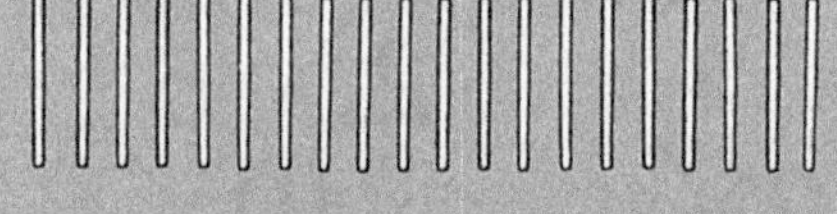

e ½ of 50

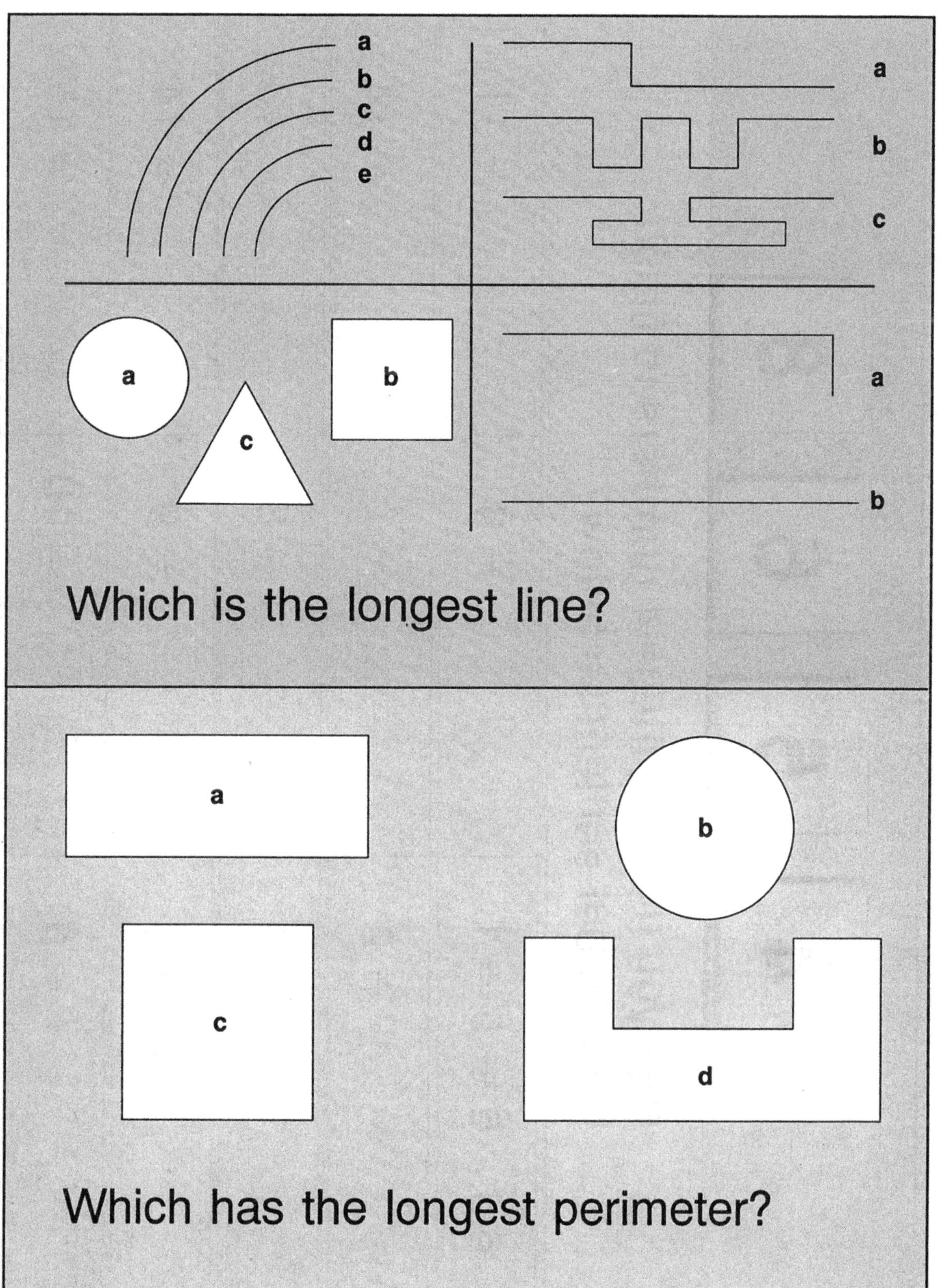
a
b
c
d
e
a
b
c
a
b
c
a
b
Which is the longest line?
a
b
c
d
Which has the longest perimeter?

4 5 6 8

Can you use just these 4 numbers to finish the number sentences below?

e.g. 6 − 5 = 1	= 6	= 11
= 2	= 7	= 12
= 3	= 8	= 13
= 4	= 9	= 14
= 5	= 10	= 15

You buy	You give	Your change
15t 3t	20t	
35t 20t	K1	
35t 40t	K1	
25t 20t	50t	
15t 60t	K1	

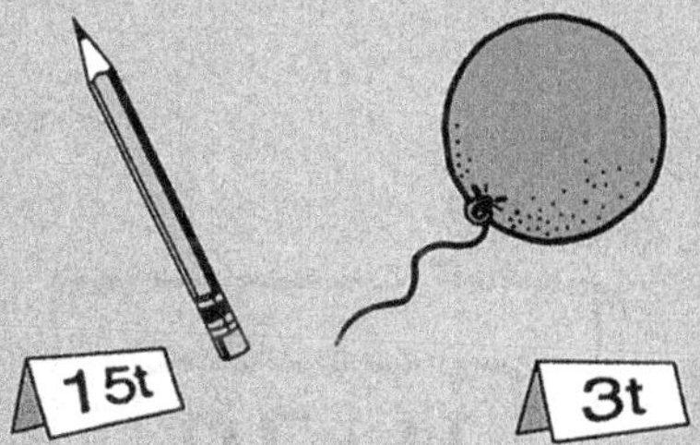

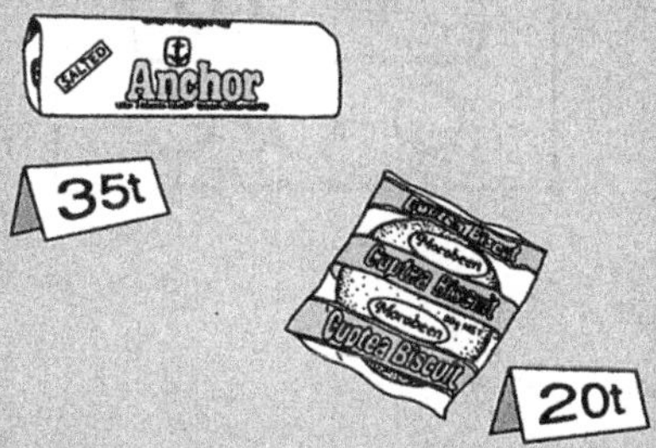

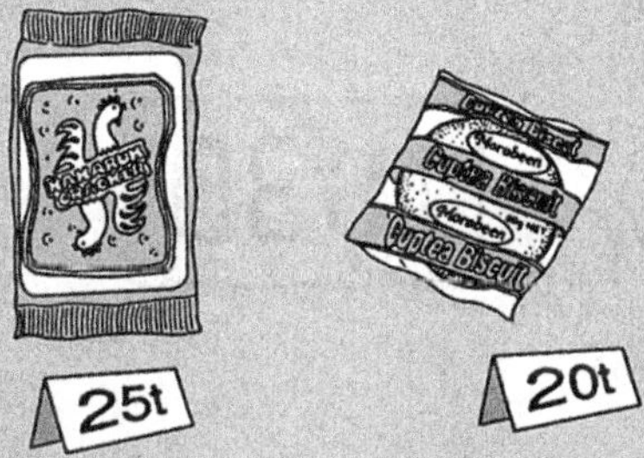

A
What is the time 2 hours before?
4:30
B
What time is it 3 hours?
11:30
8:00

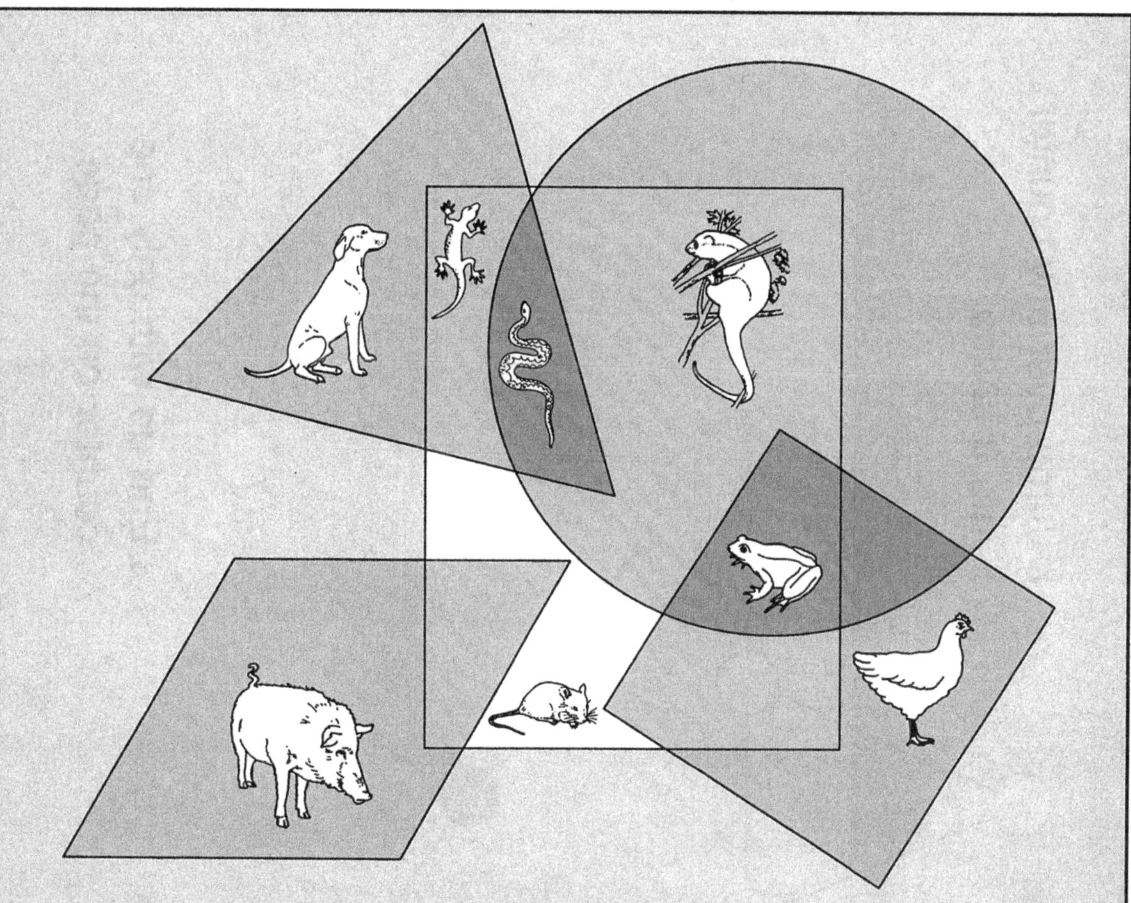

What animal is in:

a both the triangle and the circle?

b the parallelogram only?

c the rectangle only?

d the circle and the square?

e the rectangle and the circle?

f the rectangle and the circle and the triangle?

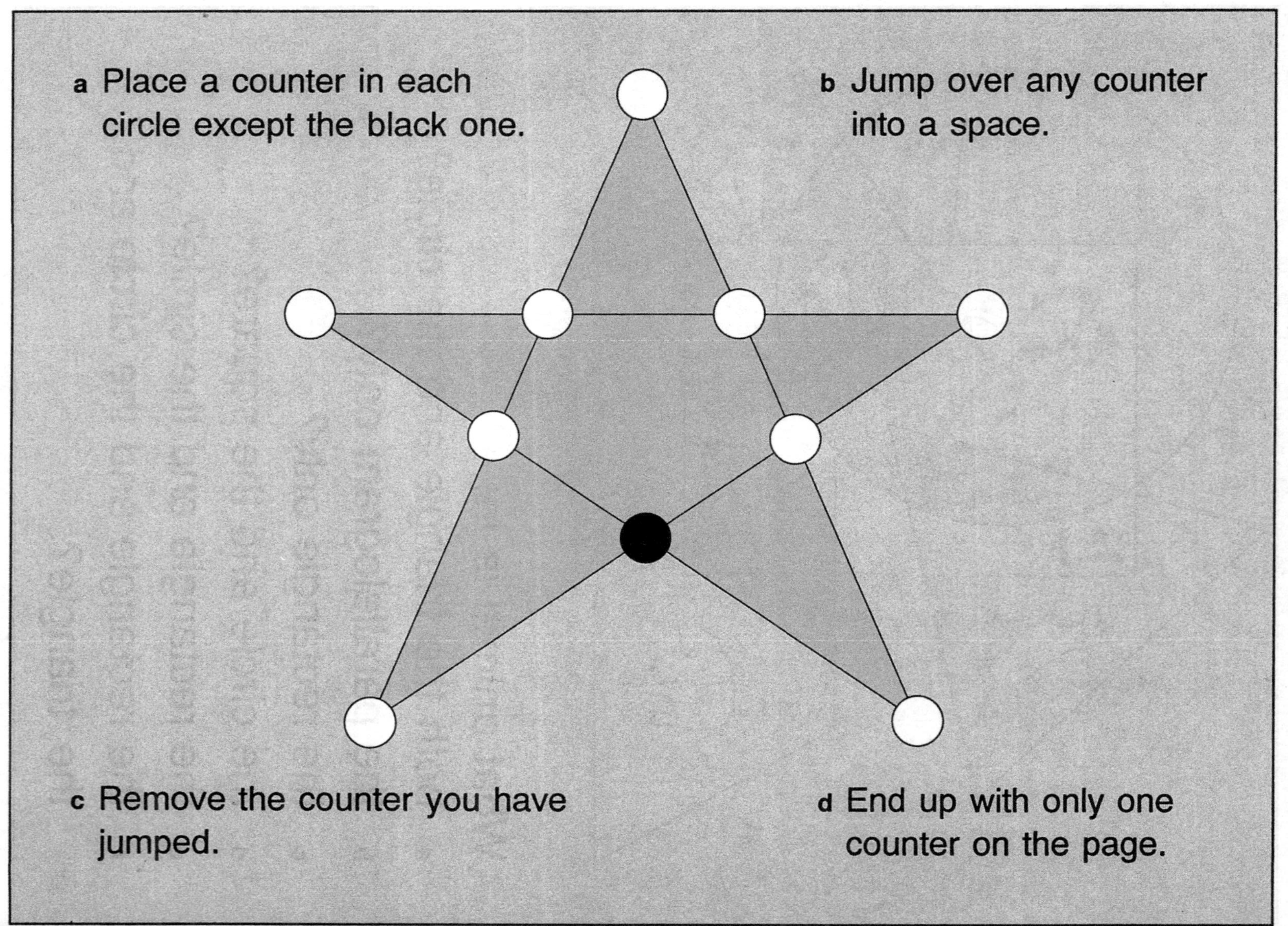
a Place a counter in each circle except the black one.
b Jump over any counter into a space.
c Remove the counter you have jumped.
d End up with only one counter on the page.